KEEPING

Bees

PAM GREGORY &
CLAIRE WARING

FOREWORD BY
PAUL PEACOCK

Publisher and Creative Director: Nick Wells
Project Editor: Catherine Taylor
Art Director: Mike Spender
Layout Design: Jane Ashley
Digital Design and Production: Chris Herbert
Picture Research: Laura Bulbeck, Joseph Kelly, Polly Prior
Proofreader: Alex Davidson
Indexer: Willliam Jack

Special thanks to: Claire Waring, Laura Bulbeck, Joseph Kelly,
Polly Prior and Chelsea Edwards.

21 20 19 18
10 9 8 7 6 5 4 3 2

This edition first published 2017 by
FLAME TREE PUBLISHING
6 Melbray Mews,
London SW6 3NS
United Kingdom

www.flametreepublishing.com

ISBN 978-1-78664-228-8

A CIP record for this book is available from the British Library
upon request.

Many thanks to the following for kindly supplying or giving
permission for image use: **Claire Waring** 8 & 90, 29t, 29b, 35,
36, 41, 42t, 43, 46, 49, 52t, 61, 64b, 66, 68, 70, 71b, 75b, 75t,
77t, 77b, 78, 79, 85, 86t, 86b, 87t, 87b, 88, 96t, 96b, 100, 102,
105, 106, 108b, 110, 111, 112, 113, 114, 115, 116, 117, 119,
124t, 124b, 125, 133, 134t, 138, 141, 142, 144, 145, 146t,
149t, 157, 158, 172b, 172t, 174, 176, 180t, 214b, 215t,

216b, 217, 219, 225t, 225b, 226, 227, 233t, 234, 237, 238b, 239, 241,
245, 249; **Tim Rowe** (www.rosebeehives.com) 9 & 80; International
Bee Research Association 18; BeeBase (Courtesy The Food and
Environment Research Agency (Fera), © Crown copyright 2015) 22b,
47, 135, 137, 143, 148, 150, 156, 161, 193, 194, 196, 197, 198, 200t,
200b, 201b, 202, 203, 204, 205, 207, 210, 213t, 213b, 214t, 220l,
220r, 228, 236; **Pam Gregory** 21, 122, 126, 136, 186; **Brian McCallum/
Urban Bees Ltd** 63; **B.J. Sherriff** (www.bjsherriff.co.uk) 67t & b; **Paynes
Southdown Bee Farms Ltd** 69, 185l; **Omlet** 81t; **Matthew O'Callaghan**
(www.warrebeehive.co.uk) 81b; **Beesource.com** 101; www.
tassotapiaries.com 134b, 211, 216t; **Maisemore Apiaries** 179, 185r.

Courtesy of Wikimedia Commons (via the Creative Commons
Attribution 3.0 Unported Licence and/or Creative Commons
Attribution-Share Alike 3.0 Unported Licence and/or GNU Free
Documentation License) and the following suppliers: **Grapetonix/
Tobias Radeskog** 19t; **WikipedianProlific/adpated by Flame Tree
Publishing Ltd** 38; **Waugsberg** 44r, 53t, 71t, 149b & 238t, 233b;
Bilby 72; **Robert Engelhardt** 166; **Luc Viatour/www.lucnix.be** 173;
James D. Ellis, University of Florida/© Bugwood.org 195; **Erbe,
Pooley: USDA, ARS, EMU** 201t; **Biovet** 206.

Courtesy of **Shutterstock.com** and © the following suppliers for
the black and white illustratrations: Epine, Abracadabraa, Maria
Letta, Neuevector; and for all other images: Cre8tive Images 1 &
15t & 42b; Eduardo Ramirez Sanchez 3 & 251; Michael Avory 4 &
33; Catherine Murray 5 b & 93, 15b & 40, 58,; 62; Darla Hallmark
5t & 57, 59; Poznyakov 6t & 129; Torsten Schon 6b & 163; Evgeni
Stefanov 7t & 191; JSseng 7b & 231; Kovalchuk Oleksandr 10;
Nayashkova Olga 11; DuleS 12r; Victor I. Makhankov 12l; Nikolay
Stefanov Dimitrov 13; LilKar 14, 34; photocami 17; Testbild 19b;
Vladislav Gajic 20; Arvind Balaraman 22t; marilyn barbone 23;
Henry Nowick 24; Elena Elisseeva 25; Hannamariah 26; And
Andreev 27; Zacarias Pereira da Mata 28; Jacinta 30; Susan
McKenzie 39; orionmystery@flickr 44l; CHRISTOPHE ROLLAND
45; darios 51, 53b, 168, 247; Rodney Mehring 52b, 242; Kirsanov
64t, 155, 232; Jens Stolt 65; S_E 73; grafvision 74; Petr Baumann
76; Tyler Olson 84; Antonio S. 94; Markus Plank 97; Agata Dorobek
98; sellingpix 99; Diane Garcia 104; Terrie L. Zeller 107, 160;
andersphoto 108t; Mudassar Ahmed Dar 109; abxyz 120; buriy
123; Max Nagornyy 130; photo25th 132; Filip Fuxa 139; Timothy R.
Nichols 140; Malgorzata Kistryn 146b; Zeljko Radojko 147; Durden
Images 152; Carsten Medom Madsen 153; clearviewstock 154;
Shebeko 159b; vitor costa 159t; Olga Miltsova 164; S.Dashkevych
169r; Alekcey 169l; Rodney Mehring 170; Aaron Amat 171;
Anisimova 175; M Pace 177; Afonkin_Y 178t; Teresa Kasprzycka
178b; Jostein Hauge 180b; carroteater 182; Anna Dlubak 183;
Michael Pettigrew 184; Mats 187; hjschneider 188; basel101658
189; suravid 192; Toth Tamas 212; Imageman 215b; mihalec 222.

KEEPING
Bees

PAM GREGORY & CLAIRE WARING

**FOREWORD BY
PAUL PEACOCK**

FLAME TREE
PUBLISHING

Contents

Man's association with bees is ancient. The art of beekeeping, or apiculture, has been honed over the millennia, and is a practice about which we are still learning. As well as providing us with a valuable foodstuff, bees are central to the food chain, pollinating the plants on which we rely for food. Modern intensive agriculture and pests and diseases are threatening the survival of the honeybee, and this chapter is full of practical advice on how to fill our gardens and outside spaces with insect-friendly plants to help these essential pollinators.

Before becoming a beekeeper, it is essential to have a good understanding of these most complex of insects. This chapter gives an insight into how the queen, drones and workers live together as a superorganism, each with a specific function within the colony. It shows how the honeybee colony follows an annual pattern, expanding and contracting in response to the available suitable food supply. To be a successful apiarist, you should be familiar with this annual pattern in order to know when you can harvest honey, when the colony requires input and when to leave well alone.

First Things First..........56

Beekeeping is extremely rewarding but also requires time, energy and a certain amount of resources. Before you fill your hive, you should be sure you are able to commit yourself to the bees, particularly during the active season. It is a good idea to join your local beekeeping association, which can help with everything from targeted local bee knowledge to the sourcing of second-hand equipment. Finally, this chapter will ensure you are ready for the less glamorous side of apiculture: the swarms, the stings, the pests and the diseases.

Your Bees 92

Once you have decided that bees are for you, this chapter will help you to set up their new home. It is important to choose your site carefully – and remember to let your neighbours know they will have new neighbours. Make sure you are familiar with all the equipment; practise using it all until you feel confident with it. Careful observation of your bees will help you get to know your new colony, and minimum interference will ensure you have a happy hive.

Through the Seasons

Apiculture follows the bees' annual cycle. This chapter gives advice on what your bees will be doing and how you should be looking after them at specific times of the year, although it should be adapted according to climatic conditions and your specific location. You will be able to use this information to plan ahead to ensure the smooth running of your hives and that you are ready for every eventuality.

Reaping your Rewards

This is the moment you have been waiting for! This chapter is all about honey: what it is, what its properties are and how the plants it is derived from affect its flavour. These pages will show you how to be sure your honey is ripe, how to extract it and how to achieve the consistency you require. And there is more to beekeeping than just honey. Beeswax can be used for candles, polish and more, so you will learn here the best way of extracting it. Lastly, learn about how to extract and make use of the lesser-known substance propolis.

Pests And Problems

Like all animals, honeybees are vulnerable to pests and diseases. Some of these can be fatal. This chapter looks at the main threats to your hive, such as Varroa, American Foul Brood, European Foul

Brood, Small Hive Beetle, and other threats including woodpeckers and rodents. It helps you to recognize a healthy colony and to know when you should take further action. You will learn how to reduce the risk to your colony by ensuring you have good apiary hygiene.

Taking Things Further..... 230

Once you have acquired the basic skills of beekeeping, you can begin to think about building on what you have learned to become a more proficient, confident apiarist. This chapter looks in more depth at various aspects of beekeeping, such as the queen and how to identify her, how to assess the characteristics of a colony and how to control swarming. The advice given in these pages is a great starting point for the beginning of a fascinating lifetime of beekeeping.

Foreword

Beekeeping has become the barometer for health in many aspects of life in the natural world and, in turn, our everyday life. We have all witnessed news reports about the plight of honeybees and may well now be familiar with words and phrases such as 'varroa' and 'colony collapse disorder'. The numbers of people hoping to help honeybees has risen to hue-and-cry proportions and local beekeeping associations are struggling to cope with training the high numbers of possible new recruits to the craft of beekeeping.

It is more than just the plight of bees driving record numbers of people to take up beekeeping. The movement towards self-sufficiency, especially in our towns and cities, is quite naturally drawing people to think about beekeeping. A hive and its associated colony add a new dimension to the garden crops available, and honey, wax and propolis are gathered by bees far beyond the borders of the vegetable plot.

Moreover, the care and synergy needed to keep bees successfully are reflected in the gardener being at one with his plot, and the desire to put bees on allotments and in gardens is a natural one. This has led to the phenomenon of the urban beekeeper, with his need for gentle bees, and a catchment of gardens in an urban setting.

There is plenty of food for bees in the countryside – brambles, clover and ivy and loads of trees being the main food sources and making fantastic honey, and there are also the pollinated crops professional beekeepers move their many hives to, to service top fruit such as apples and cherries, flax, oil-seed rape and so on. However, beekeeping has followed mankind into the cities – places where, fed by thousands of gardens, bees do well. One beekeeping friend lives by a large municipal cemetery and his bees are treated to a never-ending supply of nectar from the cut flowers of hundreds of funerals and their associated visitors.

To become a successful beekeeper you need both experience and knowledge. Ever called upon to make decisions that affect their bees, beekeepers rely on what they have read, been told by a mentor or been shown at a local beekeeping association, and their increasing experience of what actually works for them. Consequently, whatever reason they had for starting in beekeeping, their growth and ongoing success comes from a deeply held fascination with bees. They are absorbing, intelligent creatures, with much to tell us about selflessness and living in society.

Beekeeping is, at its heart, a social activity. I dare say there would not be so many apian diseases if mankind hadn't involved himself in bee's affairs. But that said, the production of food, in any context, is about people, and the majority of beekeepers are committed to their local society, usually a branch of a larger national association, where they learn new methods and, more than anything else, have a jolly good time. It is a concept I believe this book sets out well; to become a good beekeeper, you really need to enjoy what you are doing.

Paul Peacock
Self-sufficiency author and broadcaster

Introduction

Have you ever had time to watch and ponder on smaller lives than our own? Those myriad insects, for instance, that scuttle in the grass, inhabit soil and water or fill the air with their activity in drowsy, summer afternoons – have you ever wondered about their role in the scheme of things and how they affect us? This is the story of one of the most important and fascinating of those insects – the honeybee and how you can help to look after them.

The Bees

Bees are essential to life on earth. Their busy work collecting nectar and pollen ensures that plants are pollinated. This pollination service is vital for both our food and our environment. Honeybees have pollinated one mouthful of our food in every three we eat – and this applies especially to the interesting things such as fruits, nuts, seeds and oils. As well as this, honeybees ensure that wild flowers are pollinated. This is vitally important to the wild birds and mammals that need a good supply of nuts, seeds and berries to ensure their winter survival.

Save the Bees

Recently there has been a growing realization that honeybees
are in trouble and that keeping bees, as well as being a relaxing
and profitable hobby, can help to protect this most valuable
of animals – and in doing so protect our own environment, its
wealth of biodiversity and the food crops we eat.

The Products

There is nothing nicer than honey fresh from your own beehives.
Honey has long been one of mankind's most desirable foods
and, even today, for hunter-gatherer societies honey is their
only sweetener. The poet Robert Brook was left wondering 'if
there would be honey still for tea'. A successful beekeeper will
have plenty of honey for tea and for other things too. Increasing
interest is being shown in the healing properties of bee
products. Have you ever tried honey and lemon for your sore
throat? Honey can be used for many medicinal purposes – even
in hospital dressings.

Not Just Honey

Bees don't just produce honey. They also produce beeswax, which can be used for making candles,
creams and polish. Propolis, or bee glue, is a widely used antibiotic. Royal jelly is thought to have
aphrodisiac and rejuvenating effects by some (ever hopeful souls), while pollen can be used as a protein
rich food supplement. Even the venom from the sting of the bee is being researched for its potential to
relieve arthritis and multiple sclerosis.

The Beekeeper

Anyone can keep bees: young or old, male or female, of whatever colour or creed. It can be done anywhere – town or country. You don't need to have land or even a large garden. Bees can be kept in all sorts of places – even on a flat roof – and plenty of farmers and gardeners are happy to offer beekeepers a place to put their bees because they know they will reap the wider benefits. Bees in urban areas often do better than bees in rural areas because of the wide diversity of flowers available for their food.

A Certain Kind of Temperament

Beekeeping is not ideal for those who want to barge busily at life. To get the best from the bees a beekeeper needs to be gentle, observant and neat. Good stockmanship

demands you work quickly and quietly but with due care, enjoying the observation of these remarkable creatures. That is why it is so good for you – because you have to be in control, of yourself and of the bees. Being eager to start is not a bad thing but learning with other people, finding beekeeping buddies and possibly an experienced beekeeping mentor, will make the craft so much more successful and so much more fun.

12

The Benefits of Bees

Beekeeping is a fascinating hobby that holds something to interest everyone. Honey is great for eating and for cooking too. Beeswax makes candles and polish, and cosmetics and creams. For entrepreneurial beekeepers there is the chance to earn some extra income from sales of surplus produce, while for others home-produced honey or candles are perfect gifts for friends and family. And your fruit and vegetable crops will be so much better if there are bees in the garden to pollinate them.

A Boost for Wellbeing

Beekeeping is a healthy hobby. It gets you out into the fresh air and is a fantastic stress buster. Whatever your worries – the bailiffs are at the door or the spouse has run off with the milkman – you have to put them all out of your mind and concentrate on the bees to succeed at beekeeping. You need to be calm and in control of both yourself and the bees, which is excellent for reducing blood pressure and just relaxing.

Not a Lonely Hobby

The local beekeepers' association is a must, too. There you will meet enthusiasts keen to share their experiences and knowledge. Not only that, but beekeepers can have very strong views on the best ways to do things – and they don't always agree. For me this discussion is part of the fun and fascination of beekeeping.

And Much More Besides

Once you have started to master the art, you could show off your produce at local or even national honey shows. If you like gadgets and gizmos, beekeeping has plenty; while if your fancy is biology, sociology, philosophy or environmentalism the complexity of the bee's social life will offer something thought-provoking.

About This Book

This book will explain in easy steps:

▶ About keeping honeybees
▶ About their biology and social structure
▶ How to keep them safely and productively
▶ How to reap their bountiful harvest
▶ What equipment you will need to do this
▶ How to overcome some of the problems that arise
▶ Where to get more information and training

A Constantly Rewarding Hobby

People have an age-old relationship with bees and they hold a sacred place in many cultures. They are fascinating creatures with a complex social lifestyle that will leave you marvelling. After 35 years of beekeeping I feel I know less about them as each year passes – as knowledge grows, so does the realization that there is so much more still to understand. There is so much to find out and enjoy about bees that they offer a lifetime of interest – and lovely honey and better garden crops as well. What a wonderful deal we humans get from bees.

The Joy of Bees

Bees in History

For millennia people have had a close relationship with different kinds of honey-producing bees. Even today this is reflected in the way we look at bees and the words we use – 'honey' when we are talking to a loved one, we have a honeymoon after a wedding, and we use names such as Melissa, Melanie, Pamela and Deborah, which all have their origins in our relationship with honeybees.

Honey Hunting

Mankind's first contact with bees was probably through honey hunting for a sweet, delicious treat that was worth risking being stung for and it is likely that people have long recognized the medicinal value of honey. Honey hunters risked life and limb to climb up to the nests of wild bees in order to rob their honey and wax. Cave paintings bear testimony to the way it was done and similar honey hunting methods are still used in some parts of the world today. The picture here suggests that it was not only men who were honey hunters.

Did You Know?

As late as the eighteenth century, the queen bee was thought to be a 'king bee'. The idea of a queen bee in nature was contrary to the notion of the 'divine right of kings'.

Honey Beer and Mead

For thousands of years honey was one of the few available sweeteners. Later it was discovered that when honey is fermented, perhaps with spices or fruits to vary the flavour, it makes a potent alcoholic drink. Both honey beer and mead remain popular today. The gods, like Odin and Thor, in the northern sagas, drank copious quantities of it and mead is mentioned in many myths. The medieval mead-maker was a very important man with special tasks and privileges.

Refreshing sparkling mead

Ancient and Sacred Rites

Honey has long been used in ancient and sacred rites worldwide, the special value of honey relating to the sweetness of truth. Beeswax, too, was widely valued for its many uses, from providing candlelight to embalming the dead. In fact, the word

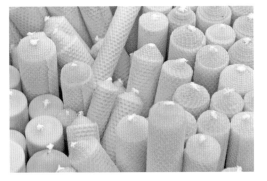

'mummy' comes from the ancient Persian word 'mum', meaning wax. Beeswax candles still have a central place in the Christian church.

Revered Creatures

Bees have also been revered by many religions as creatures that are specially blessed, and much is made of their relationship with honesty, truth and the gift of sweet words in oratory and poetry. Bees figure in many creation myths worldwide.

19

From Honey Hunting to Beekeeping

Over time, people found it was possible to improve their chances of collecting swarms of bees by attracting them into specially made receptacles rather than having to climb trees and cliffs for unpredictable rewards. This first step towards managing bees clearly established personal ownership of the resource.

Early Hives

Early hives were simple in design and constructed of freely available local materials – woven grass baskets, hollow logs, bark or clay containers. However, the downside of these hives was that the bees were usually killed when the honey was harvested. The largest and the smallest colonies were killed in the autumn, the largest because they had most honey and the smallest because they would not survive the winter. These styles of beehives are still widely used in developing countries today.

Old clay hives

Bee Space

The discovery of the bee space by Reverend Langstroth in 1852 led to a revolution in beekeeping, giving us the movable-frame hive and the beekeeping methods that we use today. Langstroth noticed that honeybees always left gaps about 7–9 mm wide around and between their combs. In fact, this is just enough space for two worker bees to pass back to back without touching each other. Spaces bigger than 9 mm are filled with honeycomb while those under 7 mm are filled with propolis, a resinous substance collected from trees such as poplars or pines.

Movable Frames

This observation led to modern movable-frame hive beekeeping and made the wholesale slaughter of the bees for their honey harvest unnecessary. Today's methods allow the beekeeper to observe and control what is going on in the beehive without disturbing the bees too much, maximizing the bee colony's chances of survival.

Bees and the Environment

People value honeybees, not only for the curiosity of their complex social system but also because healthy bee colonies are an indication of a healthy environment. Honeybees have been around for over 30 million years, over which their characteristic behaviours have evolved. These give rise to a well-ordered society, with the single, fertile queen who devotes her life to laying eggs; the busy workers who are the engine of the hive; and the drones who are pampered all summer only to be thrown out to freeze in the autumn.

Bees and Flowers

Most people know that bees live in hives and produce honey. Bees do not make honey so that we can enjoy it, although this is what happens. Likewise, plants do not produce nectar so that bees can make and store honey, although, again, this is what happens. Bees have evolved over millions of years alongside the flowers they so assiduously pollinate. This relationship is mutually beneficial. The flowers benefit from the transfer of pollen, containing the male genes, from one flower to another's stigma, containing the female genes.

Attracting Pollinators

Flowers go to great lengths to attract suitable pollinators. They have developed a whole array of colours, shapes and scents to ensure this important function is fulfilled. We benefit from this quirk of evolution by having a wonderful world of flowers and scent as well as being able to utilize the resulting fruits. The bees benefit by having a ready source of pollen and nectar.

Vital Pollination

Pollination is vital to both our food and our environment. It is essential for many of our food crops; for others, it gives higher yields, better-quality crops, or earlier results (or all three) than those that are self-pollinated. Honeybees also pollinate wild flora. This is of huge importance to the wild birds and mammals that need a good supply of nuts, seeds and berries for their winter survival. In turn, these small creatures provide food for larger ones as part of the natural food chain.

The Threats to our Bees

Recent media publicity means that people are more aware that the honeybee and other insect pollinators are in trouble; that their vital pollination services cannot be taken for granted. In 1992, a dangerous new pest, the Varroa mite, infested honeybees in the UK. This parasitic mite has crossed the species barrier from its natural host, the Asian honeybee. Unfortunately, our honeybee has no natural defences and so, for the foreseeable future, its survival depends entirely on beekeepers taking the time, trouble and expense to ensure their bees are protected against the mite.

Pests: *Braula coeca* (top) compared to *Varroa destructor* (right), *Tropilaelaps* (centre bottom) and *Melittiphis* (left)

New Threats

More recently honeybees have faced new threats: Colony Collapse Disorder, Small Hive Beetle and *Nosema ceranae*, all arising from the movement of bees and honey from place to place. In globalizing just about everything, we have become very good at globalizing the threats to our honeybees.

Reduction in Forage

To make matters worse, the amount of suitable forage available to pollinators has dropped dramatically. Over the past 40 years we have destroyed 95 per cent of our wild flower meadows, 50 per cent of our ancient woodland, 60 per cent of our heathland, 80 per cent of our downland and 50 per cent of our lowland fens. And this is just in Britain. A similar pattern is happening all over the world as increasing numbers of people use land for their immediate needs.

Did You Know?

Honeybees are one of the most useful insect pollinators because they remain constant to a particular type of flower while it produces nectar, maximizing correct pollination.

The Future

The future of the honeybee may now rest in the hands of beekeepers and others who want to help bees. Let's hope bee lovers new and old rise to the challenge so we can ensure the honeybee's survival. In fact, honeybees are so important that, legend has it, Einstein once predicted that if the honeybee became extinct, mankind would follow within four years. While it seems an unlikely prediction, it would be better not to put this idea to the test.

Gardening for Bees

In towns and villages all over Britain there are more than one million acres of private gardens. Even if only 10 per cent of these were used for wildlife habitat it would make an enormous difference to the wellbeing of a huge variety of plants and animals. And in creating diverse habitat for wildlife we are making a happier and healthier environment for ourselves.

An Insect-rich Wildlife Garden

Management is the key to an insect-rich wildlife garden. The greater the diversity, the more attractive it is to pollinators. The aim is to extend the flowering season of nectar-producing plants. The early spring bulbs (snowdrops, crocus and winter aconites) are immensely helpful to insects searching for the first pollen and nectar, while late-season heathers, golden rod, echinops, *Sedum spectabile* and

Michaelmas daisies are vital at the end of the year. Ivy provides the last nectar as the days shorten. It is a magnet for many insects well into November, producing a sugar-rich nectar and the last precious pollen of the season.

The Cottage Garden

The relaxed cottage garden is wonderful for wildlife. Many garden fruits and vegetables offer nectar as a reward for pollination. You can grow a huge range of annuals and perennials to keep your bees buzzing happily all year and many other insects and birds will also thank you for your efforts. Herbs are especially attractive to bees, and very useful too. Thyme and camomile are low growers and the stately fennel, marjoram, lavender, rosemary and lemon balm are great both in the garden and in the kitchen.

Did You Know?

It takes up to 15 years for ivy to reach the flowering stage.

Flowers All Year

Aubretia, candytuft, thrift, wallflowers, alyssum and the saxifrages make a welcome start to the year. Later the campanulas, verbascums, mignonette, bergamot, evening primrose, sweet Williams, poppies and alkanet will give a nectar-laden floral display. In disturbed ground, the old-fashioned arable 'weeds' can be grown to great effect. Flame-red poppies, corncockle, scabious, speedwell, scarlet pimpernel and cornflowers are gorgeous and make a welcome way-station for all

sorts of passing insects. For more colour and even more nectar, add Californian poppies, poached egg plants, nasturtiums and snapdragons. All the hardy geraniums and cranesbills are attractive to bees and fit well into the wild garden.

Grow Wild

If you have a hedge you can encourage some hedgerow plants. Sneezewort, stitchwort, Jack-by-the-hedge (which gives a fresh garlic flavour to spring salads), Welsh poppies, deadnettles and woundworts all make an interesting and easy-to-manage hedgerow bottom. Those stripy lawns, beloved of television advertisements, are a wildlife desert. Delight in the clover and dandelions, both essential food plants for bees. Dandelions make the most fantastic concentrated nectar that really encourages the bees as they build up their numbers in the spring, and the wax is a glorious and unmistakable yellow.

Grow from Seed

Most wild flowers are grown easily from seed, and some more common ones will just arrive with only a little encouragement. If you buy wild flower plant plugs or seeds, buy native plants from responsible suppliers. Other plants to delight the bees are purple vetch, self heal, yellow rattle, bugle, knapweed, hawkweeds and St John's wort. If your ground is inclined to be wet, try ragged robin, purple loosestrife and the lovely snakeshead fritillary.

Shrubs

If you have space, create a woodland glade. The birds will love the extra food and nesting places. Good shrubs for bees and other wildlife include hawthorn, field maple, guelder rose, crab apple, wild pear or cherry and spindle, perhaps shot though with wild rose, bramble and old man's beard. Cultivated shrubs that produce masses of nectar-sweet flowers include all varieties of cotoneasters, pyracantha, mahonia, berberis, Japanese quince and the lovely early flowering *Vibernum bodnantense*. Avoid anything with double flowers, which are generally unattractive to bees.

The Garden Pond

The garden pond is valuable to all sorts of creatures and gives bees somewhere to drink. They may even get some forage from well-placed water plants. Meadowsweet makes a beautiful and sweet-smelling foil for smaller pond plants and dramatic lime-green pollen for the bees, while native water lilies can provide shallow drinking places so the bees don't drown.

Live and Let Live

Don't be too tidy. 'Live and let live' should be the motto. Leave a small pile of logs in a corner to encourage slow worms, frogs and toads – they help to reduce pests in the garden. A garden like this will have standing room only for wildlife and you'll have the happiest bees in the area.

'Beekeeping' for Non-beekeepers

You can help bees without becoming a beekeeper. Beekeeping does not necessarily suit everyone, but we can all do our bit to aid the survival of pollinating insects. By providing sources of nectar and pollen you can enjoy the bees in your garden and know you are helping out with vital food supplies. Pollinating insects include other bees such as solitary bees and bumblebees, and all need suitable nesting sites. These can be specially made boxes or bundles of canes placed in a strategic position.

Different Bees

Not all bees are honeybees. In the UK alone there are over 200 species of bees. Wildlife-friendly gardens go a long way to help the solitary bees, the semi-social bees and our lovely bumblebees. You can help other bee species by providing them with nesting places.

Bumblebees

Bumblebees survive the winter as single, mated queens, who hibernate in stone walls, under stones or in other sheltered places. In spring they search for a suitable nesting sites. Although it is possible to buy special nest boxes, it is seldom worth the money. There are many designs of bumblebee homes but what is required is an inner box of about two litres (3½ pints) in size with an outer waterproof compartment. The entrances to the two compartments need to be offset so that light does not penetrate, and some hay or dried grass should be added to the inner compartment. Try the Bees, Wasps and Ants Recording Society (BWARS) website for more information.

Bumblebees (*Bombus hypnorum*) in a nest box

Solitary Bees

For solitary bees, such as the red mason bee, a different type of bee house is used. This is composed of bundles of hollow tubes or sticks bound together so that each bee can have its own space but still live near to its 'friends'. You can buy ready-made tubes but you can also make them for yourself. At their simplest, the hollow stalks of plants such as hogweed and bamboo are ideal. Cut the stalks into lengths about 30 cm (1 ft) long and bunch 10–20 together. Tie them neatly and hang them up in a suitable, sunny but sheltered and undisturbed spot.

A male leaf cutter bee (*Megachile willughbiella*) – a 'solitary bee'

29

Respect the Bees

In general, bees will not hurt people. They are too busy living their own lives. And while a sting may hurt a person, it is almost invariably fatal for the bee, so they do not sting unless they are really upset. However, an animal whose home is threatened will not be pleased. So learn to respect the bees, to value their contribution to our welfare and educate children to understand and appreciate how fascinating they are and how important in the cycle of life.

Bee holes in logs and cement blocks

Checklist

▶ **Ancient history:** Man's relationship with bees goes back many thousands of years. His first contact with bees was probably through honey hunting to rob nests of honey.

▶ **More than a sweetener:** Honey is used to make mead, the oldest alcoholic drink known to man.

▶ **Discovery and development:** The discovery of the bee space by Reverend Langstroth made modern beekeeping possible.

▶ **Better beekeeping:** Movable frames enable beekeepers to inspect colonies easily.

▶ **Colonies in nature:** Healthy honeybee colonies indicate a healthy environment.

▶ **Natural partners:** Bees and flowers have co-evolved over millions of years into mutually beneficial partners: bees pollinate flowers and receive nectar in return.

▶ **Vital to the food chain:** Bees pollinate wild flora, ensuring a good supply of nuts, seeds and berries for wild birds and mammals – and food for us!

▶ **Globalized threats:** Our honeybees have no natural defences against the Varroa mite and need the help of beekeepers to survive. A sharp reduction in the amount of available forage also threatens our bees' survival.

▶ **Help bees:** Planting nectar- and pollen-rich flowers is an excellent way of helping pollinators. Providing nesting places for other bees helps to ensure their survival too.

How Bees Work

Honey
ORGANIC PRODUCT

The Life of the Hive

Honeybees are social insects with many individuals living together as a single unit in order to enhance the survival of the whole. Their complex society has allowed bees to colonize the widest range of environments of almost any animal in the world. Key to this survival strategy is a strict division of labour. In effect, the honeybee colony can be regarded as a potentially immortal 'superorganism', while a single bee cannot survive on its own for long.

The Colony Lifecycle

The colony follows an annual cycle, expanding in the spring to take advantage of the abundant nectar flows and then contracting in the autumn to form a winter cluster capable of surviving cold weather. It reproduces by swarming. The original queen leaves with half the workers to make a new colony, while the remaining bees rear a new queen to continue their colony life.

The Honeybees' Home

Honeybees will make a home in any suitable cavity. A series of parallel beeswax combs hanging vertically forms the heart of the bees' home. The combs are used for raising the young bees, each in its own individual cell, and for storing supplies of honey and pollen. These combs are formed from hexagonal cells that fit together in a very efficient, space-saving design. Building this comb is one of the most remarkable feats of nature. Look at it and marvel how perfectly it is engineered to give maximum strength and efficiency.

Beeswax

Beeswax is the bees' natural nesting material. It is secreted by the worker bee from glands on the underside of her abdomen in tiny, translucent, white flakes. These are chewed until malleable and moulded into place. Workers hang together in festoons when making wax to keep the working area warm. The cells are built on both sides of a central midrib and those on either side are offset so the ends fit neatly together.

A bee secreting wax scales

The Nest Structure

The nursery, or brood nest, is always arranged in the same way, as a ball with the combs running through it. The precious brood is in the centre. Here it is protected and warm, covered by adult worker bees that maintain the developing bees at the perfect temperature (35°C/95°F). Next comes a ring of pollen surrounding the brood so it can easily be made into food for the larvae, with the honey forming a final shell around the outside of the brood nest, acting as insulation as well as food.

A Note on Pollen and Nectar

Pollen is the powder-like material produced by the flowers of plants and trees and is the male genetic material of the plant. Nectar is a sweet, sugary secretion produced by flowers to attract pollinating insects. Honeybees gather the nectar and carry it back to the hive where they concentrate it and add enzymes to make honey. Pollen is brushed off the flower onto the honeybee as it moves from flower to flower, resulting in pollination. Honeybees also use some of the pollen as well as nectar to feed the brood – the former providing the protein part of the bees' food, the latter providing the energy.

Colony Development over Time

The size of the colony grows and shrinks according to the time of year, reflecting the food resources available. Bees and plants have co-evolved to fit into cyclic, weather-driven, seasonal environments, and the bees time their activities to coincide perfectly with the plants' flowering periods.

A marked queen with her abdomen down a cell in order to lay eggs

Annual Colony Development

In winter, the bees cluster to conserve heat and energy. As the days grow longer, the queen's egg-laying rate increases. By the summer, tens of thousands of worker bees are available for foraging – collecting nectar from flowers, then processing and storing it as honey. As autumn draws in, brood rearing reduces and the colony begins to cluster ready for another winter.

The Winter Cluster

During winter, the colony consists of the minimum number of bees – perhaps as few as 5,000 workers. The queen will have stopped laying eggs, all drones have been evicted and the older workers have died. (See the next few pages for the differences between queens, workers and drones.) The colony cannot afford to carry any passengers and only the smallest possible number of the most vigorous young bees, plus the queen, will make it through the winter. The bees huddle together in a ball, clustering around the combs, regulating the temperature by moving closer together when it is cold and spreading out when it is warmer (*see* page 137).

Spring Development

As the days lengthen, the workers raise the temperature at the centre of the cluster to around 35°C (95°F). This allows the queen to start laying eggs. As the brood develops, from eggs to larvae and then pupating into adult bees, it creates heat – and as the temperature rises, more eggs are laid. As the weather continues to improve and fresh pollen and nectar become available from the early spring flowers, the brood nest expands. By late spring, the queen will be laying well and the number of workers increasing rapidly.

Maximum Colony Size

The maximum colony size coincides with the peak nectar flow. At this time, the colony will be bursting with workers, busily storing honey for the winter. Once plenty of food is available, the queen will start producing drones and the workers may even start preparing for the colony to swarm. A colony may reach a peak of 50,000 or more by mid-summer.

Did You Know?

A queen can lay more than 1,000 eggs every day.

Autumn Decline

As spring progresses into summer, the queen starts to reduce her egg-laying. The number of larvae that need feeding is reduced and workers are freed to forage for the last nectar and pollen. By mid-summer, the number of young bees hatching is starting to decline, and by early autumn almost no new worker bees are being added to the colony. The remaining bees start to prepare for the long winter ahead.

Anatomy and Life History

The honeybee colony normally consists of one queen, thousands of workers and hundreds of drones. The queen is the only individual that can lay fertilized eggs.

These develop into worker bees. Workers perform the basic tasks within the nest and fly out to forage for nectar, pollen, water and propolis. Remarkably, the queen can also lay unfertilized eggs, which develop into drones. The main function of the drone is to mate with a virgin queen, after which he dies.

Typical Insect Form

People are often surprised when they see a honeybee. They are relatively small and look more like wasps and not at all like the

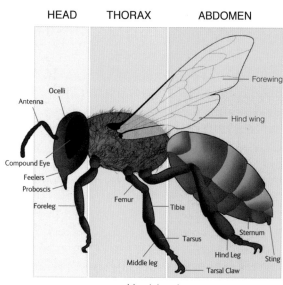

HEAD   THORAX   ABDOMEN

Ocelli
Antenna
Compound Eye
Feelers
Proboscis
Foreleg
Femur
Tibia
Middle leg
Tarsus
Hind Leg
Tarsal Claw
Sternum
Sting
Forewing
Hind wing

A female honeybee

38

more familiar fat, hairy bumblebee. Honeybee bodies have a typical insect form divided into head, thorax and abdomen. They have six legs, four wings, two compound eyes, three simple eyes (ocelli – 'ocellus' in the singular) and two antennae, richly studded with the sophisticated sensory receptors that are essential for survival and communication in highly socially adapted creatures.

Specialization

There are three specialized types of bee in a colony: one is male, known as a drone, and two different types of female, the queen and the workers. There is normally only one queen and about 500 drones, with the largest number of bees being workers. They are all governed by a rigid social structure and an age-related division of labour.

The Drones

A drone

The drones have only one function, and that is to mate with a new young queen. Because drones are designed primarily for this purpose, any physical structures related to other work are reduced while their mating-related structures are enhanced. They cannot sting and they make little other contribution to the hive, depending entirely on the provisions given to them by the workers. Indeed they are not even very faithful to one colony, roaming between hives quite frequently.

Larger Features

Drones are easily recognizable by their larger bodies, strong wings (success with passing queens goes to the fastest) and big eyes that meet on the top of their head (all the better to see the queens). They have highly developed olfactory organs (all the better to smell them!). Drones develop from unfertilized eggs by a process known as parthenogenesis.

Drone Congregation Areas

When the weather is good, drones hang about on the drone equivalent of a street corner – a drone congregation area (DCA) – waiting for young queens to come by. These aerial mating zones are usually in the same place every year and have probably developed to enhance the chances of drones and queens actually finding each other and maximizing the chances of out-crossing. Queens will mate with 10–20 drones. The act of mating kills the drone, who falls to the ground while his place is taken by the next one – but at least he dies with a smile on his face!

Did You Know?

Drones cannot sting.

Raising Drones

When colonies are prosperous and thriving, generally by late spring, the bees are willing and able to support drone production. Between 500 and 1,000 drones will be raised by the colony when they are needed. However, as 'passengers', they will only be raised during periods of plenty; they are expelled from the hive and die at the end of the season or during periods of dearth. So, drones are a measure of the health and strength of a colony and the food available.

The queen can be seen at the edge of this hive frame

The Queen

The queen bee is the mother of the colony and is essential to its survival. She carries all the heritable material of the colony so is responsible for all its physical and behavioural characteristics. Changing the queen can change the nature of the colony within a few weeks.

Egg Laying

Under most circumstances there is only one queen and she has two vitally important functions. The first and most obvious is egg-laying.

Over her long lifetime, and a queen may live for several years, she will lay thousands upon thousands of eggs.

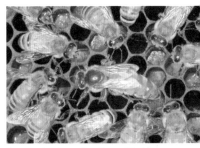

Pheromone Production

The queen's other essential role is to produce pheromones, or chemical signals, that control the behaviour of the other bees and ensure colony cohesion. Of all the pheromones used to control the communication and activity of the hive, 'queen substance' is one of the most significant.

A queen is cleaned and cared for by a group of worker bees called the 'queen's court'

Colony Odour

Each colony has its own distinctive 'odour' resulting from the pheromones the queen produces, passed around the colony as workers groom her and share food. This smell tells the workers that they are in the right hive, that the queen is present and safe, and that all is well. Exposure to queen substance contributes to the suppression of both ovary development in the workers and their interest in rearing a new queen.

Queen Features

The queen is recognizable by her long legs and longer, more pointed abdomen. Her wings only reach halfway down her back. But she is not very much bigger in size so it can be quite daunting for a novice beekeeper to find her. Fortunately this is rarely necessary. She lives her whole life where she will be most protected – inside the hive, in the dark, and at the most active centre of the brood nest. However, in an emergency, the bees can, and will, make a new queen as long as there are fertilized worker eggs or very young larvae available.

Swarming and Mating Flights

The queen will only go out of the hive once or maybe twice in her life. If the colony decides to swarm, the queen goes with the swarm to start up a new colony, leaving behind a queen cell in which will

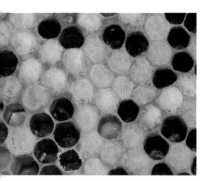

Worker cells and larger
drone cells

develop a new, young queen. The only certain time the queen will leave the hive is to mate. She does this just once in her whole life. After mating, the queen stores all the sperm she needs for the thousands of eggs she will lay in her lifetime in a small organ in her abdomen known as the spermatheca. This special organ has a valve at the opening, which allows her to fertilize an egg if she chooses. The fertilized eggs develop into workers or, under certain circumstances, a new queen, while unfertilized eggs develop into drones. It is thought she makes this choice by measuring the size of the cell with her front legs before she lays her egg in it – drone cells are larger than worker cells and this size difference can be clearly seen by the beekeeper.

What about the Workers?

The workers are the most numerous and the bees that people are most likely to see. They are females that cannot mate, so have no capacity to reproduce (but *see* page 218 in Pests and Problems regarding Laying Workers). As their name implies, they do all the work of the colony. The type of work they do is loosely related to their age, the development of their glands, the time of the year and the available food supply.

House Bees

After hatching, young bees spend a few hours finding their way about the hive and working out what jobs need doing. Then they set to work as house bees. As their feeding or brood food glands develop, they become nurse bees. Later their wax glands develop and they become honeycomb-making bees. As the colony reaches its full size, there will be 1,000

A worker bee

or more new workers emerging from their cells every day. Older workers are pushed to the edges of the brood nest where the jobs they do will change: cooling the hive by fanning their wings or receiving pollen and nectar from incoming foragers.

Guard Bees and Foragers

As their sting glands develop, the workers become guard bees and, finally, they become foraging bees collecting nectar, pollen, propolis and water to maintain the life of the colony.

Summer and Winter Bees

After only a few weeks, maybe around six or eight, the worker is worn out and will die. Foraging bees usually die away from the hive, just disappearing one day and not coming back. But while summer bees live for only a few short weeks, during the winter the worker's physiology alters and she will live for five or six months.

A guard bee

Workers' Tasks and Approximate Ages

▶ Orientating and patrolling   Day 1 onwards
▶ Cell and hive cleaning   Days 2–6
▶ Feeding brood   Days 4–12
▶ Comb building   Days 6–18
▶ Storing pollen and honey   Days 8–18
▶ Guarding the entrance   Days 18–21
▶ Foraging for food   Days 20–end

Did You Know?

The 'busy' worker bees spend a large part of the time 'resting' or looking for something to do.

Worker Anatomy

The worker has a number of anatomical features worth remarking on because they have practical consequences.

Pollen Baskets

The pair of pollen baskets on her hind legs are used to transport pollen. Stiff hairs on the inside of each leg are used to brush pollen grains from the body and pass them to the back legs. The pollen loads are very easy to see on the worker bee. It is sometimes even possible to identify the plants the bees are visiting from the colour of the pollen in their pollen baskets.

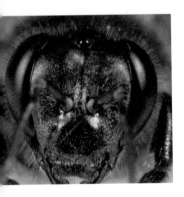

The Bee's Vision

Honeybees can see the plane of vibration of polarized light, which enables them to determine the position of the sun even if it is cloudy. As a result, bees can navigate to their food sources and communicate this information to other foragers by their dances, even in the darkness of the hive. The many lenses that make up their compound eyes give a blurry picture, so they don't easily see shapes but movement is detected very quickly. This is why people should not flap about and wave their arms to ward off a bee – this will attract the bee's attention as an aggressor. The best thing to do is to keep still and the bee(s) will be less aware of you.

The Sting

Only female bees can sting. The worker's sting is the only means of colony defence. The barbs on the sting enable it to penetrate into the

A bee's stinger

aggressor and deliver a dose of venom. However, once the worker has used her sting the barbs prevent her from withdrawing it, so the sting mechanism tears from her body and she dies.

The Queen's Sting

The queen also has a sting but it is smooth so she does not get damaged if she uses it. The queen's sting is only used to fight with a rival queen if more than one queen is present in the same hive. This only happens when a number of virgin queens are reared, normally as part of the swarming process. They then fight to the death until only one remains to take over the colony, or the first queen to hatch will kill the others while they are still in their queen cells.

The Busy Bee

In the summer, when there are plenty of flowers and the weather is hot, the workers (foragers) can collect as much as 12 kg (25 lb) of honey in just a few days. In fact, these foragers are little 'honey-collecting automatons'. They can't help themselves. If the weather is warm enough for them to fly and there is nectar to be had, they will gather it and turn it into honey. As a result, they often collect far more than they need. It is some of this surplus, of course, that the beekeeper harvests.

Enviable Fuel Consumption

Each colony, which can have up to 50,000 bees in the summer, will collect around 57 kg (120 lb) of honey to see it safely through the year. It takes a trip equivalent to three orbits of the earth to produce one pound of honey. As 30 g (1 oz) of fuel for each 'earth orbit', the flying worker bee is running at the equivalent of 1.5 million miles to the litre (7 million miles to the gallon). Additionally, up to one litre of water is needed each day for cooling the hive, producing larval food and diluting honey, amounting to up to 50 trips for the specialized water collectors.

Life Cycle

Individual bees go through a four-stage development, starting as an egg that develops into a larva. The larva then pupates and turns into a fully formed adult worker, drone or queen. Each of these has a different overall development time, with the queen being the shortest and the drone the longest.

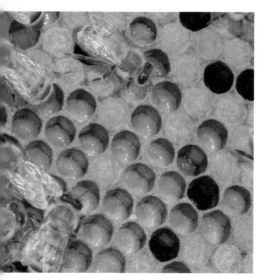

Larvae inside their cells

The Egg

The life history of the individual bee starts with the queen laying a single egg in a cell of the comb. The house bees clean out and polish the cells before the queen lays a single egg in each. The egg hatches after three days.

The Larva

The new larva is fed by the nurse bees, first on royal jelly and then on a mixture of modified royal jelly and pollen to create a new food called bee bread. The larvae grow fast for five days, then, when they have grown to their full size, their cells will be covered over, or 'capped', with wax so that they can pupate in private.

The Pupa

Pupation is the miraculous process in which the larva turns into a fully formed adult (or 'imago'). It takes just 21 days for an egg to become an adult worker bee. Drones take longer to develop because they are bigger – in all, 24 days from egg to hatching. Surprisingly, queens develop much faster, taking only 16 days from egg to a new queen emerging. Development times can be compared in the table below.

A hatching bee

Brood Development Times

Days	Queen	Worker	Drone
0	egg laid	egg laid	egg laid
1			
2			
3	hatches	hatches	hatches
4			
5		diet changed	diet changed
6			
7			
8	cell sealed		
9		cell sealed	
10			cell sealed
11	5th moult (prepupa)		
12			
13		5th moult (prepupa)	
14			5th moult (prepupa)
15	final moult		

Continued overleaf

47

Days	Queen	Worker	Drone
16	emerges		
17			
18			
19			
20	mature	final moult	
21		emerges	
22			
23			final moult
24			emerges
25	mating period begins*		
26			
27			
28			
29		flies	
30	begins to lay		
31	end of mating period*		
32			
33			
34			
35		mature	
36			
37			mature
38			
39			
40			
41	too old to mate	foraging begins	

* marks the period of time during which the queen can take her mating flight

Behaviour

Inside the hive it is hot, dark, crowded, smelly and electrically charged. In such a complex environment, excellent communication between all members of the colony is essential. There are four main methods of sharing information – food sharing, pheromone signals, dancing and vibrations.

Food Sharing

Food sharing is the continual offering and accepting of food between bees, spreading certain types of information very rapidly through the colony. One worker will beg for food and another will offer it, regurgitating it from her honey stomach. While they are sharing food, they are also touching each other and passing on pheromones. This touching and food sharing allows bees to recognize each other. It also tells other workers about the quality of the incoming nectar, what types of nectar are most in demand and whether more water is needed.

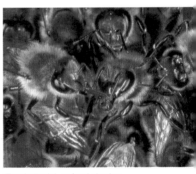

'Trophallaxis', or food sharing

Pheromones

Scientists think as many as 39 pheromones may be needed to control all the colony's activities but only about 17 have so far been identified. Pheromones don't just tell the workers that the queen is safe.

49

Alarm pheromones tell other workers to come and help defend the colony from predators (or beekeepers), while the Nasonov pheromone is used for marking things so it helps a lost bee get home or a swarm of bees to a recognize its new nest site.

Communication Dances

Dances can tell the bees where the food is located and how abundant it is. They can encourage the bees to swarm and direct the swarm to the new nest site, and they may help to protect a virgin queen or give her away to her emerging rival queens. Three basic dances have been identified, although others have also been noticed. The food dances are the best understood.

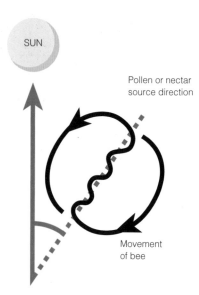

SUN

Pollen or nectar source direction

Movement of bee

The Waggle Dance

The most sophisticated is the waggle dance, which indicates a food source located some distance from the hive. This dance is performed on the comb face. The vertical direction indicates the exact position of the sun. The dancing bee runs in a figure of eight with the centre of the wagtail line angled to indicate the direction of the food source relative to the sun. The rate and duration of the dance and the number of figures of eight completed all relate to the distance of the food source while the vigour of the dance indicates the richness of the food source.

Colony Reproduction

Swarming is the act of colony reproduction. This process is entirely natural but does not necessarily occur every year. A reduction in the level of pheromone produced by the queen may be one swarming trigger. Another is believed to be congestion in the hive, which makes it more difficult for all bees to receive the queen's pheromone signal.

Colony Reproduction

If plenty of food is available then the colony will build up a strong force of worker bees. A strong and prosperous colony of bees may eventually decide to swarm. Swarming is the colony's natural form of reproduction and it is probably one of the most spectacular events in the life of a colony. However, bees do not necessarily swarm every year.

Queen Substance

Bees are thought to swarm in response to the level of the queen's pheromones. As long as there is plenty of queen substance from a vigorous queen, the instinct to replace her will be suppressed. The quantity of pheromones the queen produces depends on her age, whether or not she is effectively mated, the time of day and the time of year. Distribution of the queen substance depends on how crowded the hive is. In queens more than about 18 months old, the strength of their pheromones starts to reduce, starting the process of her replacement – this can be either by supersedure or by swarming. We will talk about swarming first because it is the most common, and also the most difficult for the beekeeper to manage.

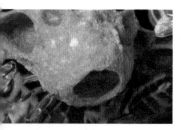

A queen cup

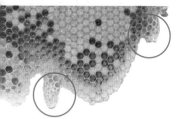

Two queen cells can be seen along the edge of this comb

Queen Cups and Queen Cells

Once the concentration or dispersal of the queen substance becomes reduced or impeded, for whatever reason, the workers' queen-cell-building behaviour will cease to be suppressed and they will start to make swarm preparations. Once this has been initiated, a well-defined sequence of events takes place. The first is construction of queen cups. These look like acorn cups and do not become queen cells until they contain an egg or larva. We will talk more about queen cups and cells and how the beekeeper can recognize them when we come to look at swarm control methods.

Royal Jelly

The queen lays fertilized eggs in the queen cups and the workers feed the developing larvae lavishly with royal jelly, a special food produced by the brood food glands of young worker bees. This rich, high-protein food fed to the queen larvae (there will be more than one and may be many) means they develop very rapidly into virgin queens.

Slimming Down the Queen

As the queen cells develop, the workers feed the old queen less and less. This means she stops laying eggs and her abdomen shrinks dramatically. At the same time, the workers 'shake' the queen frequently, again to keep her moving and reduce her weight. This slimming down process is so that the queen will be better able to fly.

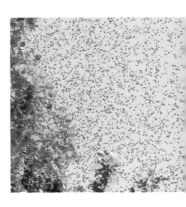

The Prime Swarm

As soon as the first queen cells are sealed, the first (or prime) swarm will take to the air. When a new swarm issues, the air is full of thousands of excited, flying bees searching for a place to cluster and trying to keep track of the queen. About half the worker bees in the colony will leave, together with the old queen.

The Swarm Cluster

After flying a short distance, the swarm clusters together temporarily in a suitable place while scout bees look for a new nest cavity. These scouts tell the other bees about the best nest sites by dancing on the surface of the swarm cluster, using the same language as the food dances, and they will keep investigating and communicating nest site information to the colony until a democratic consensus is reached about which place is best.

The Virgin Queens

The parent hive is essentially queenless for some time after the prime swarm has left. It may take a week or more before the first new virgin queen emerges from her queen cell.

After-swarms or Casts

If the colony is still strong enough it may issue smaller after-swarms, or 'casts', headed by young, unmated queens, and this may continue until the colony is no longer able to stand further splitting. If the colony does not send out casts, then the first queen to emerge will kill all the other virgin queens.

The New Queen

After around five days the young queen will be ready for mating, and ideally will take her mating flight in the next two weeks. This lasts for no more than 20 minutes and she will make a maximum of two mating flights.

Advantages and Disadvantages of Swarming

The advantage of swarming is that the old queen gets a lot of help from the workers in establishing a new nest site. The disadvantage is that it takes a great deal of energy investment and exposes both the old and the new sections of the colony to great risk. If the forage is insufficient or the weather conditions become adverse, both parts of the colony may perish. The loss of swarms severely depresses honey production as a significant portion of the foraging population of the colony leaves to form the new colony.

Supersedure

Supersedure is queen replacement without swarming. Older or injured queens are at most risk of supersedure, and larger colonies more likely to supersede than smaller ones. In a perfect supersedure, the old queen remains laying eggs until the new queen is mated and laying, after which time she may either politely die or be killed by the workers. During an imperfect supersedure, the old queen may die before the new queen is developed. The disadvantage is that, if the new queen gets lost on her mating flight or fails to mate successfully, a new queen cannot be raised because of the lack of young brood produced by the old queen.

Checklist

▶ **Honeybees are social insects:** They live together in a colony consisting of a queen, drones and workers, which together can be regarded as a superorganism.

▶ **Colony lifecycle:** The colony follows an annual pattern, expanding and contracting in response to the available forage throughout the year.

▶ **Maximum colony size:** At its peak, a colony may contain around 50,000 worker bees.

▶ **The queen – the colony mother:** The queen is the only one in the colony to lay fertilized eggs. She is recognizable by her long legs, longer more pointed abdomen and relatively shorter wings. Except to mate or to swarm, she remains inside the hive all her life, at the most active centre of the brood nest.

▶ **Communication through chemicals:** Pheromones produced by the queen (queen substance) control the behaviour of the other bees.

▶ **Drones:** Male honeybees (drones) are only present in the colony during the active season. Their main purpose is to mate with new virgin queens, after which they die. A queen may mate with 10–20 drones.

▶ **Workers:** These are the most numerous bees in the colony and perform a number of different tasks within the colony before flying out to forage for nectar, pollen, water and propolis. Worker honeybees sting but die when they do so.

▶ **Communication through dance:** Worker bees perform dances on the vertical comb to communicate the distance and direction from the hive of a nectar source.

▶ **Making new colonies:** A colony reproduces by swarming.

First Things First

Are Bees for Me?

Beekeeping may not be for you, however passionate you are about nature. It is a very practical skill and takes time to learn. Previously, beekeeping knowledge was passed from parents to children because everyone kept bees. Traditional learning routes have all but vanished. You will need more than just a good book to guide you through. Reading about it is not the same as being confronted by an open hive containing a large colony of bees.

Taster Days and Courses

Ideally, a prospective beekeeper will look for one of the dozens of taster days and novice beekeeping courses to make a bit of sense of the whole thing. Many local beekeeping associations throughout the country run such courses. Ideally, check out the tutors' qualifications. Have they passed any beekeeping exams? Do they have any teaching qualifications? Does the course have a practical element or is it all theory? You will probably get a better course that way and a good tutor is more likely to be able to explain the 'Why?' as well as the 'How?'

Learning Together

I think it is nicest to learn with another person. When I started beekeeping, I worked with a friend and we helped with each other's bees. She noticed things that I did not and vice versa, so we both learned better. Beekeeping takes time to learn and needs careful observation combined with growing understanding. Knowledge comes gradually and with practice, so don't be discouraged.

Your Local Beekeeping Association

Top Tip

You will make mistakes. The important thing is to learn from them. For this reason, it is wise to have more than one colony as soon as possible. With a second colony, mistakes and disasters can be more easily rectified.

I really advise you to join your local beekeeping association as well as attending the course. Here you will not only be able to learn from visiting lecturers and attend apiary meetings to learn more about opening and handling bees, but you will also find a group of like-minded people only too willing to give help and advice. As a beginner, you are very likely to need this support.

Other Benefits of Membership

Membership of the local beekeepers' association may offer you insurance against the risk of losing your bees to a notifiable disease. It may also be possible to take out third-party insurance (in case your bees cause problems to others).

Varying Advice

A word of warning: it is a beekeeping truism that there are as many ways of doing things as there are beekeepers, who may not necessarily agree with each other! This can be very confusing. I suggest you find a beekeeper whose way of doing things you like and stick to that advice until you have sufficient knowledge and experience to be able to discriminate the useful from the less so.

Top Tip

The most trustworthy information sources are government bee inspectors. Their services vary from country to country but those in England and Wales (from the Food and Environment Research Agency, or Fera) are particularly useful.

Finding Information and Support

There is a wealth of information about beekeeping. Unfortunately it ranges from the bonkers to the old-fashioned with all shades of helpful between these extremes, and it is very hard for the new beekeeper to unravel what is useful. The Internet is the worst place for an indiscriminate plethora of information, so treat it with care.

Trustworthy Sources

In England and Wales there is a completely free beekeeping inspection and advisory service that is absolutely top class. Very experienced beekeeping inspectors are available to help keep colonies disease-free and give advice on general husbandry. In most other places there are also beekeeping advisors, but they are not always quite so easily available and not always free. Find out what services are available in your area, and also what legal obligations you are under.

Beekeeping Magazines

There are several national beekeeping magazines that provide articles dealing with a wide range of beekeeping subjects. Your local beekeeping association will probably have a newsletter including local information and contacts.

National Beekeeping Associations

As well as local associations there are national associations These offer leaflets, more advanced courses, beekeeping examinations and national events such as honey shows and conventions. These offer the highest-quality, best-informed and latest beekeeping information. If your local association is affiliated to one of the UK national associations, you may be covered automatically for third-party insurance.

Start Conventionally

Basically, I suggest veering away from extreme ideas, at least to start with. Until you are comfortable with what you are doing and understand it all, I suggest you stick with what is most commonly done in your area. Conventional hives are easily sold if you want to move into something less conventional at a later date, and conventional methodology is well researched and well understood, and has stood the test of time.

The Time Factor

In general, beekeeping is not a hugely time-consuming hobby. Indeed, it is better that the bees are not disturbed unnecessarily. However, they do require a regular time commitment throughout the spring and summer. In the late spring and early summer you may need to inspect the bees on a weekly basis, while at other times of year the demands are rather less. If you are not able to make this commitment then it is best not to undertake beekeeping.

Commitment

I have heard it said that the difference between a good gardener and a bad one is a fortnight. This is true of bees also. The bees will do things in their own time – not yours – and it is really easy to neglect a colony. They are not like dogs, cats or other pets that demand immediate attention, and the bother of dressing up and lighting the smoker can mean those important little jobs are postponed. However, the bees will not wait for you.

A hive on a rooftop of the Lloyds building, London – urban bee hives may need unusual locations

Other Worries

You don't need to be Superman to keep bees, although supers (the boxes that are used to contain the honey harvest – placed on top of the brood box, usually above a queen excluder) can be heavy when they are full of honey – at least beekeepers always hope they will be. Beekeepers are as diverse as the general population and I have known plenty of elderly individuals, tiny women, people with disabilities and even children (under adult supervision) who are very enthusiastic beekeepers. The physical aspects can be overcome by the sensible use of equipment or setting up the hives and the apiary in ways that make life easier.

Space Requirements

You do not need to be a large landowner to have bees. A private place in a reasonably sized garden will be fine. However, remember to check out what the neighbours think. Some allotment associations allow bees (to the benefit of all allotment holders), while others have by-laws preventing it.

Out Apiaries

Bees can be kept in an 'out apiary'. This is an apiary that is not at your home. With so much publicity about bees, there are many people willing to host colonies in their garden or on their land. Your local association can often help here or you could advertise for a suitable site.

Practical Skills

Finally, you do not have to be a skilled carpenter to be a successful beekeeper. Although carpentry and beekeeping go nicely together and it is lovely to make your own equipment, there is a well-established equipment supply industry out there. Everything you need, and more, is easily, if not always cheaply, available. Some things, such as frames (the parts of a hive that carry the comb) and beeswax foundation (starter sheets of beeswax), are relatively inexpensive and it is really not worth making them unless you have very special reasons for doing so.

Stings

Getting stung is something that is very worrying to new beekeepers. The one thing most people know about bees is that they can pack a punch with their tails, but it is this sting that ensures they are treated with respect by most animals. This sting has guaranteed the bees' survival since early cave dwellers first braved the danger to rob the sweet honey from their nests. Remember also that once the worker stings the sting mechanism tears from her body and she dies, so she is not all that keen to sting anyone.

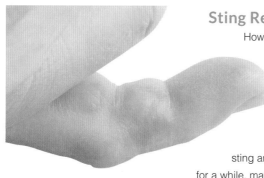

Sting Reaction

However, let us look at what happens, because most people will get a few stings and find them painful for a few seconds when it first happens – although a stinging nettle is worse in my opinion. The main reaction occurs afterwards when the body reacts to the foreign protein that has been injected. The sting area will swell up and be uncomfortable and itchy for a while, maybe even a couple of days, before disappearing.

Sting Relief

Proprietary antihistamine creams for insect bites can be helpful, especially if they contain a local anaesthetic and cooling agent, or the homeopathic Apis mel will help to relieve the symptoms. However, most beekeepers do not generally need to use anything. If you are worried about your personal reaction to stings, it may be reassuring to seek medical advice before starting beekeeping.

Immunity

Most beekeepers gradually gain an immunity to the swelling after keeping bees for a few seasons. In about 10 per cent of cases, people have a stronger reaction and swell up, sometimes alarmingly. The speed of such systemic reactions gives a rough indication of their seriousness.

Allergic Reaction

In very rare cases, a hypersensitive person can be allergic to even a single bee's sting. This can cause a rash, palpitations, breathing difficulties and unconsciousness, and requires urgent medical intervention. However, the normal outcome is a quick recovery after treatment. The skill, of course, is to avoid getting stung in the first place.

Getting Started

When you have completed the course and decided to go ahead, what happens next? You will need personal equipment such as protective clothing, gloves, a smoker and a hive tool. Your bees will need a hive and you will need to decide which design you wish to use.

Buying Beekeeping Equipment

Beekeeping is 'equipment rich' and it is difficult to know at the start what is needed and what isn't. As a beginner, it is best to stick initially to the conventional route used and taught locally. That way, more people will be able to help you if you get stuck and you are more likely to be able to buy things second-hand.

A selection of smokers

Local Equipment Sales

Your local association may have an annual bee sale or members may be able to tell you the best places to buy or know people who are giving up and have things to sell. Commercial beekeeping companies can supply everything you need, but at a price. Buy as little equipment as possible until you have a clear idea of what you need.

Second-hand Equipment

Buying second-hand is a good way to control costs but you must clean and sterilize second-hand equipment to avoid bringing any disease into your apiary. Hive boxes can be scorched with a blowtorch, while other items can be soaked in a strong washing soda solution and/or scrubbed well.

Essential Equipment for Starters

▶ Good protective clothing

▶ A good smoker

▶ A hive tool

▶ A beehive plus bees (ideally two colonies)

▶ Three or four supers per hive (*see* page 85)

▶ A spare hive, with frames of foundation or drawn comb

▶ A feeder for each hive (*see* page 123)

Protective Clothing

The most important thing is that you enjoy your beekeeping, feeling in control of the bees and comfortable when you are handling them. The first things you need are protective clothing and a smoker. My advice is to get the best you can afford.

The Veil

A veil is absolutely essential to protect the face and no one should ever inspect a honeybee colony without one. They come in various designs but the main design essentials are that you feel safe in it and that you can see though the mesh. The simplest and cheapest set-up is a tube of black net stitched onto a wired cloth hat that keeps the veil away from the face. These 'free standing' veils are normally worn with an ordinary cotton boiler suit. Care has to be taken to make sure bees cannot get inside.

All-in-one Suit

My own preference is the 'all in one' suit and veil. The original 'Sherriff' style has a hood with the black mesh veil at the front supported by two rigid semi-circular arches to keep the mesh away from the face, and a zip enabling the hood to be thrown back when not required.

Gloves and Boots

Gloves are recommended to protect the hands from stings, especially when you are just starting. Even later on, gloves will help to keep hands clean for those delicate jobs, such as queen handling, that need bare hands. The best gloves are probably a pair of rubber washing-up gloves. They are easy to wash and keep clean and disease-free. Wear these with elasticated cuffs to cover any gaps between the sleeves of your bee suit and the gloves. Other available styles are often clumsy and cumbersome. Most people wear Wellington boots to protect the ankles. They also give you a good foot grip so that you do not slip.

The Smoker

The smoker (*see* picture on page 66) is the second absolutely vital piece of beekeeping kit. It is used to control the bees in a beehive. You should never open a colony without having a lighted smoker available. No one really knows why smoke controls bees, but it does. Some say it simulates a forest fire and triggers an instinctive reaction to gorge on honey – necessary before fleeing from the nest to find another safer home elsewhere – and when the bees are full of honey they are less likely to sting. The smoker is a simple but effective design. A nozzle at the top of the firebox directs the smoke and a grid at the bottom keeps the air intake clear. This, plus a set of spring-loaded bellows, allows a draught to be forced up through the fuel.

A standard hive tool

A J tool

The Hive Tool

The hive tool is another 'must-have' piece of equipment. There are various designs, materials and makes, and people have their own preferences. I like a stainless steel hive tool because it is easy to keep clean. Broadly, there are two types of hive tool: the 'standard' and the 'J' tool. The main difference is that the curve at the end of the 'J' style hive tools is designed to insert neatly under the ends (or lugs) of the frames and lever them upwards.

Top Tip
Never put your hive tool down during an inspection.

Spare Equipment

I have already said it is a good idea to have more than one colony. However, with the costs of bees at the moment, the likelihood is that starting will most probably be with a single nucleus colony (or 'nuc'). This is a small colony consisting of a queen and a few thousand workers, plus frames of brood and stores that will build up over a season into its full size. Ideally though, you will plan for a minimum of two colonies. You do not need to buy the second colony because you can divide your original colony once it gets big enough and you have gained the skill necessary to do this. For each full colony, as well as the hive, you

69

will need three or four honey supers complete with frames and foundation to give a strong colony enough working and storage space. However, a nucleus is unlikely to need more than one super in the first year.

Room for a Swarm

As your apiary grows, try to keep a spare hive complete with frames, ideally of drawn comb (*see* page 71). Then, should a swarm occur – either from your own hive or from someone else's – you will have the space to house it. Remember, the bees won't wait for you. If you are not ready for it, the opportunity to increase your colonies may be lost. Each hive will also need a feeder. We'll go into much more detail about the choice and variety of feeders in the next chapter.

Frames

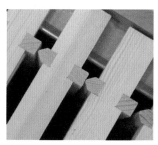

Before we get into the nitty-gritty of beehives we will look at the frames. These are the working part of the hive so they are very important. They fit inside the hive boxes and support the comb, where all the colony activities are carried out. They are designed to be lifted in and out of the hives and moved between hives, so the bees can be flexibly managed.

Hoffman frames showing spacing

Self-spacing Frames

There is a variety of frame designs but my suggestion is to go for Hoffman frames, which have shaped sides that keep them the correct distance apart. Frames normally come in two sizes – deep ones (which fit in the brood box) and shallow ones (designed to fit into the supers). This makes the supers lighter to lift.

Foundation

Frames are generally fitted with a sheet of beeswax imprinted with the hexagonal cell pattern, called foundation. This encourages bees to draw out the cells from the pattern – and they will remodel some of the wax to save them some of the enormous amount of energy it takes to make wax from their bodies. Foundation is normally wired to give it greater strength so the frames can be preserved from year to year.

Drawn Comb

Drawn comb has been 'drawn out' into honey comb by the bees – usually from the sheets of foundation. Bees use five or six pounds (2.2–2.7 kg) of honey to produce a pound (450 g) of wax. So the use of foundation gives the bees a helping hand. Many people do not like to buy second-hand drawn comb because it presents a disease risk. However, it isn't always easy to get comb drawn out (the bees just muck about with the foundation if the weather conditions aren't right for them) so drawn comb is really precious – especially if you are new to beekeeping.

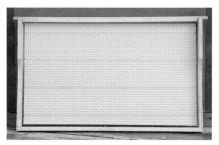

Beeswax foundation in a frame

Drawn comb

What You Do Not Need

▶ **Fancy leather gloves:** They look lovely but they restrict your dexterity and are a nightmare to keep clean. Eventually they will go stiff and you will throw them away. In the meantime they offer a risk of spreading disease between colonies. Be brave and stick with the washing-up gloves.

▶ **An extractor and extracting equipment:** You do not need these until you are closer to your first honey crop. In your first season you can probably borrow one; after this you will have a much better idea of what is available locally and how large an extractor you will need.

▶ **Anything to do with queen rearing:** This can wait until you are more established as a beekeeper.

An Introduction to Beehives

Beehives come in several designs, essentially only differing in their size and whether the bee space in the modular boxes is above or below the frames. The choice of hive is down to the individual but it is worth investigating the most popular one locally as this is probably best suits the local strain of bees. The internal dimensions are important and these must be correct to maintain the bee space throughout the hive.

Natural Nest Sites

Honeybees naturally live in suitable cavities, such as in a hollow tree, and with our beehives, of whatever type, we are trying to replicate this situation. The ideal natural nest site for honeybees is a minimum of two cubic feet in size, dark, dry and easy to defend. Bees do not mind where

they live if it complies with these conditions. I have seen them in post boxes, washing machines, dustbins, church roofs and chimneys.

The Beehive

A beehive is basically a box with one or two design features that make it convenient for the beekeeper as well as nice for the bees. The bee space is the precise gap that the bees require to move comfortably around their nest and they keep it free from any obstructions.

73

Movable Frames

Following the discovery of the bee space, bees could be kept on frames, containing comb, that were not fixed to the body of the hive. These frames can be easily removed from the hive, inspected and then replaced without damaging the bees. Any disturbance is very limited and the bees quickly settle down again. The beehives we use today are mostly modular and simple in design, meet the bees' needs and are easy to use.

Do Not Mix Hive Types

It is very important to stick to one size and type of beehive, otherwise things will not fit together. This will be frustrating and irritating, and may not be good for the bees. There is a confusing range of slightly varying beehive standards and designs. This is hardly surprising given how much beekeepers like to experiment and innovate. Someone will have a new idea and encourage others to follow it. A new approach may have some improved features or it may be worse than the conventional way. Unless you are very experienced, it can be difficult to decide what is best.

Top Tip

Reject any offers of beehive parts that are not compatible with your chosen design. If you have a mixture of designs, you will never be able to find a part that fits properly when you need it urgently.

74

Decide Which Hive to Use

You need to find out a little about beehives and decide which type you prefer. Once you have decided, stick to it so that all the frames will fit all the boxes, all the boxes will marry up with bee-tight joints and they will all be compatible with the floor, inner cover and roof. I recommend that you use the hive type that is most popular in your area.

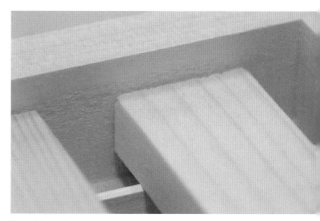

Box with top bee space

Hive Designs

The majority of beehive designs only differ in the volume of the brood and super boxes and whether the bee spaces between the modular sections (or boxes) are arranged at the top or the bottom of the box.

Top or Bottom Bee Space

There is much debate as to whether top bee space or bottom bee space hives are better. I must confess I don't think it matters very much. As long as there is a bee space between the boxes, the arrangement is the same as far as

Box with bottom bee space

the bees are concerned. However, if you mix hives with top and bottom spaces, the bee space will be violated. At one join you will have two bee spaces and the other no space at all. National, WBC and Commercial beehives have a bottom bee space, while the Langstroth, Dadant and Smith hives are top-bee-space designs. How to tell the difference? Look at your hives closely with the frames inside. The beehive has a bottom bee space if the tops of the frames are flush with the top of the box they are in. If they are not flush but are set down a bit, it is a top-bee-spaced hive and the bottom bars are level with the bottom of the box. In effect, the bee space is either above or below the frames.

Top Tip

The internal dimensions of a hive must preserve the bee space around all parts of the hive. The hive must be 'bee-tight' so that the only way in is through the entrance.

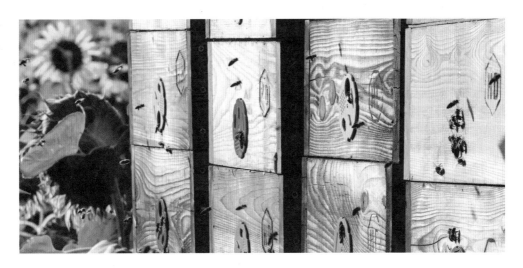

Double-brood hive

Brood-and-a-half hive

Choosing Hive Size

Different hive designs may utilize frames of different sizes and these will have different numbers of cells in the comb within. In warmer areas of the country and/or those with more prolific bees, you may want to consider one of the larger hives to accommodate your colony, while in colder areas you will probably need something smaller.

Double Brood and Brood-and-a-half

The modular nature of beehives makes it possible to alter the amount of space for brood (eggs, larvae and pupae) in the hive: you can use two brood boxes or one brood box and a super, one on top of the other (termed 'double-brood' or 'brood-and-a-half' respectively), with the queen having access to both. Your choice will depend on the weather and forage conditions in your locality, another reason for talking to local beekeepers about popular hive choices in your area. If you follow suit, supplies are likely to be more readily available and there should be a buoyant second-hand market. Generally I would recommend using the Modified National hive if you live in the UK and the Langstroth hive elsewhere.

77

Hive Types

Understanding the differences in the size and shape of different beehives is important when making your hive choice. It may be a bit tedious having to learn the features and the differences of different hives, but this will pay off in the long term and give you a much more enjoyable beekeeping experience.

A WBC hive

The WBC

The hive most obviously different from the others is the WBC (named after William Broughton Carr). This is an attractive-looking hive, perhaps a 'typical' beehive familiar from stereotypical country pastiche pictures. It is rather a small hive with an inner set of boxes covered by outer covers (or lifts). The advantage of this double-walled arrangement is that it gives some extra weather protection but, in general, it may be a little too small and the external lifts make it expensive and cumbersome to use. The frames fit into National hives.

Single-walled Hives

The National, Smith, Commercial, Langstroth and Dadant hives all look very similar, especially when seen in isolation. The boxes are designed to be used as they are with no outer cover and they are known as single-walled hives. However, Commercial, Langstroth and Dadant hives are quite a bit larger than the National and the Smith hive. This is inclined to make them heavy to use, if this is a consideration, and in cooler, wetter areas, the colony may not get large enough to fill the hive comfortably, perhaps leaving it exposed to cold and damp. These designs are not interchangeable.

Smith and National Hives

The Smith hive is similar in size to the National and they are easily confused at a cursory glance. The frames are the same size but the lugs of the Smith frame (the bits that support the frame in the hive) are smaller. This means that the Smith frame will fit into the National hive, but the National frame will not fit into the Smith hive unless you shorten the lugs.

National hives

The Rose hive

Other Types of Hives

Recently there has been a good deal of discussion of other hive types – the Rose, the Warré, the top-bar and the Beehaus hives being the most talked about.

The Rose Hive

The newest and probably the best of this bunch is the Rose hive. This is a modular hive similar to the others described, the key difference being that the brood and super boxes are all the same size, which means a different management system is possible. No queen excluder is used; the queen and the rest of the colony move freely throughout the hive, arranging the broodnest as they choose. Extra boxes are added wherever they are needed by the bees. Proponents argue that this is a more natural way of beekeeping and leads to healthier and more productive colonies. It is also a very inexpensive hive because of the way it is made. The disadvantage is that supers are heavier than those of a National. The frames are not interchangeable with any other hive type, although the boxes will fit a National or a Commercial hive. Also, the chances of spreading disease are increased by using the same boxes for both honey and brood – especially in parts of the UK where European Foul Brood is a high risk.

The Top-bar Hive

The top-bar hive has the advantage of being easy to construct. It is a horizontal hive and the design and installation make it easier on the beekeeper's back but it ignores the bees' innate preference for expanding upwards rather than outwards. It is large and rather tricky to manage, and the comb attachments are fragile so need handling with care.

The Beehaus hive

The Beehaus

The Beehaus (a derivative of the Darlington hive) is also a horizontal hive but made of plastic. Again it is large but it does have frames that are interchangeable with those in the National hive. One disadvantage is that plastic is not a comfortable material for beehives. Bees are always better off in wooden hives because of the complex water balance demands of the colony. Plastic is also very difficult to sterilize properly. However, the Beehaus can be colour-coordinated with the beekeeper's beesuit, which may appeal to those preferring style over substance.

The Warré Hive

The Warré hive was invented in the eighteenth century. Although it is a top-bar hive, in practice it is not really a movable-frame hive. The frames are fragile and comb easily sticks to the sides of the boxes. This means it is not open to any kind of modern management. Some people consider it to be a more 'natural' way of keeping bees with better ventilation. If you are not curious to know more about your bees and don't mind if they are damaged should they have to be opened for any reason, this may be a hive to try.

Top Tip
Understand how to manage your bees before experimenting with unconventional hive designs.

Warré hives

81

How Hives Work

Having looked at the various hives on offer, let us look at how they work. I remember when I first started learning about beekeeping I found all this stuff really hard to understand (and very tedious, so my apologies). I did not really see what it all meant until I saw a beehive in real life. So do try and do this as soon as possible, even if it is empty.

The Floor

This is the bottom of the hive. It will be placed on some kind of hive stand to give a comfortable working height, stop damp and discourage creatures from getting into the hive. With the advent of the Varroa mite, floors have been changed from the more old-fashioned solid ones to ones that incorporate a mesh panel and a removable tray. This allows the population of mites to be monitored and controlled. Mesh floors should be the first choice of the new beekeeper and this should be checked when buying your new hives.

Floor Construction

The floor has raised edges on three sides, forming an entrance on the fourth when the brood box is placed on top. The entrance can be reduced in size with a removable entrance block so the colony can defend itself more easily when it needs to, and to keep mice and other pests out during the winter. If the weather is very hot, the entrance block can be removed to improve ventilation of the hive and speed up the bees' access.

Parts of Hives

▶ The floor

▶ The brood box

▶ The queen excluder

▶ The super(s)

▶ The inner cover or crownboard

▶ The roof

Did You Know?

A top-bar hive is constructed rather differently from the more common style shown here. Much simpler in make-up and less precise, its body tends to be a long horizontal box, covered by a series of 'top bars' laid side by side, from which the bees construct hanging comb. Though these hives use fewer materials and are more flexible in design, care must be taken with the more fragile comb.

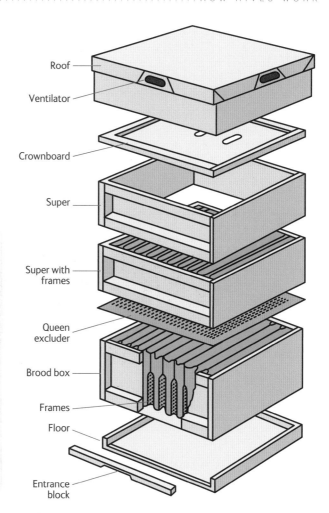

Roof

Ventilator

Crownboard

Super

Super with frames

Queen excluder

Brood box

Frames

Floor

Entrance block

The Brood Box

This stands on the floor and is where the main activity of the colony takes place. It is called the brood box because it is where the queen lays her eggs, so it therefore contains the developing brood. It also contains pollen and honey to feed the developing larvae.

Brood Box Construction

The brood box is a deep box that contains 11 frames in most hive designs. The frames are supported by their lugs, which rest on a small ledge on either side of the brood box. To prevent the frames from being stuck down with propolis, a narrow strip of metal or plastic runs along the length of the ledge. Remember, it is by examining how the frames fit that you can check for top or bottom bee space.

The Queen Excluder

The queen excluder is a framed grid or a sheet of slotted metal or plastic the same dimensions as the cross-section of the hive. It sits across the top of the brood box and keeps the brood part of the hive separate from the honey storage area, usually a super. Worker bees can pass through the gaps into the honey storage areas but the larger drones and queen cannot, so their activity is confined to the brood box. There are different types of queen excluder to choose from, but my own choice is the wire excluders (called Waldron or Herzog excluders), which are gentler on the bees' legs and wings when they pass through and give greater ventilation to the hive.

The Supers

During the active season, foragers bring nectar back to the hive and store it as honey. The colony's natural inclination is to store honey above the brood nest. To give them space for this, additional boxes, called supers, are placed on top of the brood box extending the nest upwards.

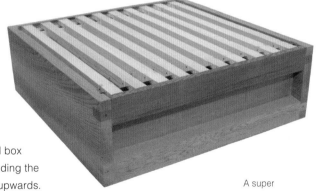

A super

They are fitted with frames in the same way as the brood box. Supers have the same cross-section as the brood box but are generally shallower in height. They take shallower frames to fit their height dimensions. The advantage of this is that the supers are lighter to lift when they are full of honey.

Top Tip

As a general rule, three or four supers are needed for each colony.

85

The Crownboard

This is a flat board, or inner cover, that
fits neatly onto the top of the brood box
or supers to close off the hive. It usually
doubles up as a clearer board, used at
harvesting time to clear bees out of the
supers, and has one or two holes cut into it for
feeding the colony.

A box fitted
with a crownboard

The Roof

The hive is completed with a watertight roof. This has
ventilation holes at the sides to allow air to flow
through the hive. These should be covered
with mesh to prevent wasps or robber
bees gaining access.

Warm Way and
Cold Way

Before we leave this discussion of beehives I will
briefly mention the idea of 'warm way' and 'cold way'. This
is only applicable to a square hive because only these allow a
choice of orientations of the frames to the hive entrance. The concept

The roof

Frames placed in the 'warm way'

Frames placed in the 'cold way'

is quite simple. When the brood box is placed on the floor with the frames running parallel to the entrance, this is the warm way. When the frames are placed at right angles to the entrance, this is the cold way. These arrangements allow you to alter the airflow through the hive.

Which Way?

Personally, I always use the 'warm way' which has a number of advantages, the main one being that when you are inspecting the hive, you are standing behind it, keeping out of the way of the activity at the entrance. The brood nest is roughly spherical with the combs slicing through it. With frames the warm way, the brood nest area stays in the middle. If the frames are placed in the hive 'cold way', the brood nest is generally found towards the entrance side of the frames with the pollen above and honey stored towards the back of the hive.

87

Acquiring and Choosing Bees

There are several ways of acquiring your first colony and this should not be rushed. Many new beekeepers are, quite naturally, desperately keen to get their first colony of bees, which sometimes leads to unsuitable stock being purchased. There are four main choices: purchase a nucleus of bees from a reliable equipment supplier, purchase a nucleus from someone in your local beekeeping association, purchase a full colony from someone giving up beekeeping, or collect a swarm (*see* page 116).

Buying Bees

If you decide to buy, the first thing you need is a guarantee they are free of major bee diseases. There are four notifiable bee diseases: American Foul Brood (AFB), European Foul Brood (EFB), Small Hive Beetle (SHB) and the *Tropilaelaps* mite. If you know or suspect that your bees are infected by any of these, you are duty bound by law to report this to the authorities. SHB and *Tropilaelaps* have not yet been found in the UK and we hope that this situation will continue.

Transferring a nucleus to a brood box

Nucleus Colonies

The most sensible choice is to start with a nucleus colony from a local beekeeper. Your nucleus should be ready to expand into a full-sized colony by the end of the season in which it is purchased.

Top Tip
Wherever you get the bees from, ask your local Bee Inspector or someone equally knowledgeable to inspect them.

Criteria for Nuclei

Ask the seller if the nucleus complies with these criteria (laid down in the British Beekeepers Association information leaflet B14: *The BBKA Standard Nucleus Guidance Notes*) or similar criteria likely to be stipulated by your country of residence:

- The nucleus should contain bees, brood, food and a queen of the current or previous season
- At least half the total comb area should contain brood in all stages (eggs, larvae and sealed brood) with at least 30 per cent being sealed
- The frames should be well covered with bees
- The frames should be securely pinned/nailed together and be sound; they do not need to be new
- Combs should be fully drawn, not foundation
- There should be enough food to last the colony at least two weeks without any further food intake
- There should be no active queen cells at any stage of development
- The bees should be good tempered when handled by a competent beekeeper in suitable conditions
- The brood and adult bees should show no signs of disease
- The seller should tell you what treatments, including for Varroa, have been given and when they were administered. There is a legal requirement to keep such records. You should check all treatments given are legal, as once you purchase the bees, you become responsible for any illegal substance that may be found in the honey or wax

Different Types of Bees

Bees belong to the genus Apis, which is part of the order Hymenoptera, and our bee is the Western honeybee, *Apis mellifera* (literally the honey bearer). There are four major subspecies of this honeybee:

The Dark European Honeybee

Although it has been widely hybridized with imports of more prolific European races, the Dark European honeybee (*Apis mellifera mellifera*) is the native honeybee of the British Isles. These bees are well adapted to temperate climates, overwintering well with good disease resistance and are 'thrifty' with their stores.

Continental European Races

The continental European races, such as the Italian (*Apis mellifera ligustica*), the Carniolan (*Apis mellifera carnica*) and the more minor Caucasian (*Apis mellifera caucasica*), are often very gentle bees, but do not always thrive in a temperate climate, being best adapted to cold winters and reliably hot summers. However, if these conditions apply, they will generally produce more honey than the Dark European. The most widely used bees in the US are Italians (the majority) and Carniolans.

Hybrids

A large proportion of the bees in the British Isles are hybridized mixes of all four European subspecies because of the large volume of imports, especially the 'Italian' type, from New Zealand and previously from the USA. While this out-crossing can produce hybrid vigour, in bees it can also lead to bad temper.

Gentle Local Bees

Probably the best choice is to find local bees that are gentle and easy to handle. These will be reasonably well adapted to survive in the local area and have stable genetics so that they are likely to stay that way. Ideally, the queen will be home-produced.

The Dark European honeybee

Checklist

▶ **Is beekeeping for you?:** Are you ready to commit? Beekeeping is a practical skill which needs to be learnt and you must be prepared to spend time with your bees during the active season (spring to early autumn).

▶ **Be aware of the risks:** You are likely to get stung and should be prepared for possible reactions.

▶ **Ways to get started:** It is useful to attend a course for beginners before acquiring bees. Find an experienced mentor to help you, and join your local beekeeping association.

▶ **Personal equipment:** Acquire a good veil, gloves and boots, and choose a smoker and hive tool that suits you.

▶ **Buy your frames:** Choose which frames you want – Hoffman being a good, easy-to-work-with choice.

▶ **Many designs to choose from:** Choose a hive design and stick to it. This way you will be able to easily swicth around and replace parts.

▶ **Learn your way round your hive:** Become familiar with the different parts of the hive.

▶ **The bee space:** Ensure all parts of the hive are spaced according to the'bee space' – i.e. 7–9 mm – and check hives with top and bottom bee spaces are not mixed up.

▶ **Keep your hive clean:** Sterilize any second-hand equipment you acquire.

▶ **Be ready to receive swarms:** Make sure you have spare equipment.

▶ **Know where your bees come from:** Take care when acquiring your first colony of bees, making sure it meets the criteria laid down by the beekeeping authorities.

Your
Bees

Practical Beekeeping

Having looked at what you will need and where to get it, it is time to start looking at the practical aspects of beekeeping. You need to choose and arrange your apiary, assemble your equipment, learn to light your smoker and keep it going, transfer your nucleus to a full-sized hive, handle bees, maybe catch a swarm and feed your bees for the winter.

Choosing an Apiary Site

Bees choose their nest sites using criteria that have been developed over millennia of evolution. These include the volume of the cavity, the degree of shelter and the orientation of the entrance. Bees can be

kept in many different places, but siting beehives is a decision that can affect the bees, the beekeeper and other people in the area, so it needs to be made carefully. While in general bees can be kept anywhere, some locations will be better than others.

The Perfect Site

It will never be possible to find the perfect place, but take as many as possible of the following into consideration.

▶ Sufficient space for working and storage

▶ Easy access (supers can be heavy)

▶ Diverse food sources (forage) within easy reach

▶ No nuisance posed to neighbours and the public

▶ Sheltered, especially from east winds in winter

▶ Not in a damp, cold or very exposed spot or a frost pocket

▶ Protected from animals and vandals

▶ Near an available source of water

▶ Find an out apiary if your home site is unsuitable

Top Tip

Make sure that you have sufficient space in your apiary to put down hive parts during inspections.

Forage

Bees need a diverse range of bee-friendly plants within their flying range, some 2–3 km from the apiary. A good diversity of plants enables bees to be kept in one place all year round and helps make sure there is enough pollen and nectar outside the main foraging season. Ideally, there should not be too many other stocks of bees nearby that might lead to excessive competition for forage.

Water

All living creatures need water and bees are no different. They use water to cool their hive and rear their brood. Scout bees will search for a suitable source and then return and recruit other foragers, so the

water source needs to be fairly close to the hive if they are not
to become a nuisance at neighbouring ponds and paddling
pools in the summer. A shallow container filled with pebbles
just covered with water will allow the bees to drink without
drowning, and the water will warm up quickly on a sunny day.
Make the bees' drinking place a feature of a garden pond so it
is attractive as well as functional.

Accessibility and Practicality

Beehives need a relatively level position and both bees and
beekeeper need sufficient space. The beekeeper needs
enough space to work with the bees comfortably. If colonies
are too close together, the disturbance of opening one hive
will upset the adjacent one, making it much less pleasant, if
not impossible, to work with the bees.

Space Between Hives

As a rule of thumb, allow nine times its footprint per hive –
that is, if you think of the hive as taking up one square, you
need the equivalent of one square to the left, one to the right
and two behind it. This will be reduced if you keep hives
in pairs. You need room to work, and especially to stand
behind the hive when you have it open. You will also need
space to each side and behind so that you can put down
the roof out of your way when you are inspecting the colony.

Sufficient Space

Make sure your apiary site is large enough to accommodate the maximum number of colonies you plan to keep, as well as the spare equipment that might need to be stored there. Ideally, no more than ten stocks of bees should be kept permanently in one apiary or they will suffer from competition for food.

Transport Considerations

Honey is heavy to carry so think about the practicality of transporting a good crop of honey from the site. It is also very nice if you can find a place for a small, preferably dedicated, storage shed close at hand, to house the inevitable clutter of equipment that you will need. There is nothing worse than missing something vital once you have the colony open.

Protection from Adverse Weather Conditions

Your apiary needs to provide the bees with the best conditions for working. Colonies need protection from extreme weather conditions, especially those biting north-east winds in winter. Bees blown down by cold winds will never recover enough to get back to the hive and are lost to the colony. A bee whose body temperature falls to 9–10°C (48–50°F) cannot move sufficiently to raise that temperature and will die.

Hedges

A thick hedge will slow down the wind speed along the ground for 40 times the height of the hedge. Choose a place with good air drainage, avoiding frost pockets and very exposed or excessively wet places.

The Sun's Warmth

If the weather is cool, colonies will benefit if their hives are exposed to the sun. The sun's warmth on a hive can make a huge difference to the capacity of the colony to start work early and to fly out to defecate in winter. However, it is also possible for hives to overheat in very hot weather, leading large colonies to start swarming preparations or, in extreme cases, for the honeycomb to collapse.

Ideally, your apiary will not be near busy, public places

Neighbours

Not everyone is as fond of your bees as you are. If you are on good terms with your neighbours, talk to them before you get your bees. Explain that you will need to open the colonies weekly in the summer, especially if your work pattern is such that this will mostly be at weekends.

Siting the Apiary

Avoid busy public places. Arrange the hives so that the flight paths of the emerging bees avoid people. A hedge or bushes in front of the hives can make the bees fly upwards away from people. Often, if people cannot see the hives, they will be unaware that the bees are there.

Swarm Nuisance

A swarm may cluster anywhere, and this could be in your neighbour's garden. It might even decide to take up residence in the next-door chimney or cavity wall. For neighbourly harmony it is important to prevent or control swarming. If there is a hint of trouble, move the bees as soon as possible. Happy relationships are of even greater value than bees. Better to 'bee safe' than sorry.

Out Apiaries

Bees do not need daily attention so they can be kept in an out apiary. This could be in a farmer's field, an orchard or any spare piece of land. You do not need a big area. It is, however, absolutely essential to keep livestock away from the hives.

Rent

You can offer to pay the rent with honey. The traditional reward is a pound of honey annually for each beehive. You will have to travel to an out apiary, and will therefore need some kind of access for a vehicle. Moving everything in and out with a wheelbarrow or carrying it is simply not practical in the long run.

Apiary Design

The arrangement of your hives within your apiary can be important. In a small apiary, hives can be placed side-by-side (0.6–0.9 metre/2–3 feet apart), facing the same direction. Longer rows can lead to

the bees drifting into the wrong hives, so larger numbers of hives
and stands should be arranged in a less uniform manner.

Facing South

Given the choice, bees will choose a place with the entrance
facing south. In practice this may not be possible, and it is fine
to orientate hives to suit the situation as long as the apiary site is
nicely sheltered.

Hive Stands

Hives need to be raised off the
ground on some kind of stand.
Stands need to be sturdy because
a hive with supers full of honey can
weigh over 100 kg (220 lb). Your
hives should be level, although
some beekeepers tilt theirs
forward slightly so any water or
condensation can run out. There
are lots of different hive stand
designs. One of the simplest is two
9-inch concrete blocks per hive.
Alternatively, you can construct a
neat purpose-built frame on four
sturdy legs.

Assembling Equipment

You can purchase hives and frames assembled and ready for immediate use, but this is an expensive way of getting kitted out. Most beekeepers purchase hive and frame parts in the flat and put them together themselves when they are required.

Making Your Own

You do not have to be a carpenter to be a beekeeper. It is possible, perhaps even desirable, to buy everything you need. Although you can buy everything already assembled, this is an expensive choice. If you are desperate to make your own hives, plans for the WBC, Langstroth and other hive parts can be found at www.beesource.com/build-it-yourself/, or the BBKA sells plans for the WBC, Modified National, Smith and Langstroth hives (*see* www.britishbee.org.uk).

You can download plans for hives from websites such as www.beesource.com

Top Tip

Before you nail your hive boxes or frames together, check they are square and frames are not twisted.

101

Flat-pack Kits

The most popular option is to buy hives and frames as flat-pack kits. They are easy to assemble and offer nothing like the frustration of putting together flat-pack furniture. A new DVD takes you through the construction of a flat-pack hive (*see* www.bee-craft.com/shop). In general these kits are really good value for money and, because they are machine-made, they fit together very accurately.

Suitable Timber

Your hive will take a lot of knocks during its life so it needs to be made from timber at least 19 mm (½ inch) and preferably 22 mm (⅞ inch) thick. Hives are now generally made of pine, which is very durable, particularly if you treat it with a preservative. However, this *must not* contain insecticides.

Foundation

You will also need beeswax foundation to put into the frames. Remember, frames and foundation come in two sizes: deep for the brood box and shallow for the supers. Both are inexpensive relative to the work it takes to make them yourself. The foundation will normally have wires in to give it strength in use and for extracting. Other types of foundation are used for more specialist purposes, such as extra-thin and unwired foundation for cut-comb honey.

Stocks of Frames and Foundation

It is good to have sufficient packs of foundation (stored flat, at room temperature) and frames in hand to be able to make up new frames in advance of them being needed, say for a swarm. They will need putting together. However, don't keep too much wax in stock as it gets brittle and less attractive to the bees.

Welcome Home

You have prepared your apiary and equipment and have acquired your bees. Now it is time to bring them home. You may have to collect your new colony, so you need to know how to transport it safely. Buying locally helps to minimize the stress placed on the bees when travelling. A nucleus may come in a standard nucleus hive, a travelling box or a non-returnable temporary container. Make sure you understand what equipment, if any, needs to be returned or replaced with equivalent new items such as frames and foundation.

Transporting Bees

In general, colonies are moved in the evening when the bees have stopped flying for the day. When moving bees, they must have enough ventilation so they do not overheat and suffocate.

Ventilation

Unless the bees are already in a special travelling box, the inner cover and roof should be replaced by a travelling screen incorporating a ventilated, bee-proof mesh. This should be done in advance of the move so the bees can settle down again. The hive must be strapped securely together so that the parts cannot move apart. Special hive straps are available from equipment suppliers. The entrance must be securely closed. This closure can be as simple as a suitable length of foam. The bees must not get too hot, so they may need a gentle spray of water onto the travelling screen at intervals. This is a generally only a danger if the journey is long.

For Further Comfort and Safety

When travelling with bees in the car, it is sensible to wear your veil (the Sherriff-style veils unzip allowing them to be thrown back for convenience). Position the hive securely so it cannot be dislodged during the journey and, if possible, orient the combs in the direction of travel.

Top Tip

Ensure your bees are secure but have sufficient ventilation during transport.

At the Apiary

At the apiary, place the colony in its final position immediately. Open the entrance to let the bees out and replace the roof. The bees will orientate and learn the position of their new home.

Introducing bees to their new home

Homing Instinct

The position where the hive is placed in the apiary will be the colony's home for a long period. The bees know exactly where they live (much as we do with our own homes) and will continue to return there, even if the hive is moved. This means that a new bee colony moved into the apiary must be placed in its permanent position straight away.

Moving Bees

If the bees need to be moved, the general rule of thumb is that they must be moved less than three feet or more than three miles. Otherwise the flying bees will continue to return to the original hive location and then become lost. The loss of many flying bees in this way can be detrimental and even cause the colony to die out. During the winter this rule is not so important, as long as the weather is cold enough to ensure the bees are confined to the hive when they are moved.

Transferring your Nucleus

Eventually your nucleus will be large enough to be moved into a full-sized hive. Here's how:

▶ Prepare the full-sized hive and make sure it is complete
▶ Prepare sufficient frames fitted with foundation (or drawn combs if they are available) to make the frame numbers up to the full quota (usually 11)
▶ Move the nucleus box and its contents to one side
▶ Put the new hive in its place with the entrance facing the same direction
▶ Gently smoke the nucleus entrance to calm the bees
▶ Remove the nucleus roof and the inner cover and gently puff smoke over the top bars
▶ Use your hive tool to lift the first comb gently out of the box
▶ Gently remove the frame, trying not to roll or crush any bees, and transfer it and all its adhering bees into the new brood box; the first frame will be the most awkward
▶ Move all the rest of the frames, and their bees, across, transferring them and placing them in the same order, centred in the larger brood box
▶ Fill up the new brood box with the prepared frames
▶ Replace the crownboard and roof

Keeping Safe

Bees do not sting unless they are annoyed. Unfortunately, lots of things annoy bees. They hate dark colours, which is why beekeepers traditionally wear white clothing. They also hate woolly, rough or textured materials because their feet get caught in them, upsetting them and causing them to sting. This also applies to hair, so beware. Working with honeybees should be a pleasurable and enriching experience. However, it is important to remember that bees can be dangerous and must be treated with respect.

Avoid Strong Smells

Keeping clothing clean helps to reduce stings because the smell of previous visits, and especially any unnoticed stings, remains on the clothing and upsets the bees at the next visit. Bees hate the smell of human sweat. Watchstraps trap sweat and bees will often sting there. Avoid using scented soaps and eating strong-smelling food, such as garlic, before visiting the bees.

Top Tip
If you get a rash or start itching all over the body, feel nauseous or faint, have difficulty breathing or heart palpitations, seek urgent medical help.

Bee Safe

▶ Think of others when siting and opening hives

▶ Handle bees gently; avoid banging or bumping them – they hate it

▶ Try to open the bees on 'good' beekeeping days (if you feel comfortable in a T-shirt it is probably perfect weather)

▶ Never open hives in the late evening or at night

▶ Have good equipment and protective clothing

▶ Let someone know where you are and how long you expect to be

▶ Get stung in company before undertaking colony inspections on your own

▶ Have the means (e.g. a mobile phone) to call urgent medical help if you show signs of a violent sting reaction (anaphalaxis)

Controlling Bees with Smoke

One of the skills of beekeeping is to use smoke to keep bees under control when opening the hive. As a novice you should use plenty of smoke. The skilled beekeeper will need only minimal smoke and know exactly when to apply it, having learned to judge the bees' temper.

Smoke Protects the Beekeeper

Smoke also protects the beekeeper while working. When a bee stings, it releases an alarm pheromone that other bees recognize as a danger signal and they will come to add their stings at the same site. Consequently, it is important if you get even one sting that you cover the place very quickly with smoke to mask the smell.

Getting Smokers Going

Lighting the smoker and keeping it alight can be one of the more frustrating aspects of beekeeping. The real test after lighting it is to put the smoker to one side and leave it. If it is properly lit it should still be capable of producing smoke 30 minutes later. Practise beforehand so you get it right before you open your colony.

Smoker Fuel

The fuel you use is important. You should use only natural materials, but these are many and various. The most important thing is that the fuel gives cool smoke that will last for the whole time you are working with the bees. In general, fuels that are easiest to light will produce cool smoke but burn away quickly and often deposit lots of tar in the smoker. Fuels that are harder to light often produce hotter smoke but last longer and deposit less tar. In general, a mixture is best. Whatever smoker fuel you use, make sure that it is dry.

Suitable Fuels

- ▶ Hessian sacking
- ▶ Corrugated cardboard (or egg boxes)
- ▶ Sisal string
- ▶ Grass clippings or dried Leylandii leaves
- ▶ Wood shavings
- ▶ Rotten wood
- ▶ Pinecones

Top Tip

Don't forget to put the smoker
out safely after use.

Start with Easy Lighting Fuel

Personally, I scavenge for fuel. I generally start with egg boxes, because they are easy to light, then, when they are burning well, I add other fuels that last longer and produce less tar. I like the smell of pinecones and I sometimes add herbs I have dried. If you use sacking, make sure it hasn't been used to store anything that may give off noxious fumes; with cardboard, check it has not been treated with a fire retardant. Interestingly, partly burned fuel lights more easily the next time and a larger smoker is easier to keep going than a smaller one.

Using the Smoker

Use the smoker at specific points while opening and handling to keep the bees under control. Bees always react most anxiously to the first cracks of light as you lever the boxes apart. It is therefore very important to apply smoke quickly at the place where the hive is first opened. Once it is fully opened, the bees will settle, especially when smoke is applied over the tops of the frames. Gentle smoke should be drifted over the tops of the frames to keep the bees inside the hive. Don't be afraid to apply smoke when you are working with bees. It will not harm them and, as you get more experience, you will need to use it less and less.

Cover Cloths

Using cover cloths reduces the need for smoke, as most of the hive remains in darkness. These are simply two cotton cloths of sufficient size to cover all but the small gap needed to work with one frame. Commercial ones are weighted at each side to help hold them on top of the hive, but tea towels will do the job. However, they can increase the risk of spreading disease between colonies if they are not kept very clean.

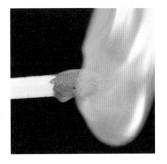

Handling the Bees

A good beekeeper works quickly, quietly and purposefully with the bees. Be gentle when handling bees. They hate being knocked or banged. They also interpret rapid or jerky movements as a threat. Be clear about the work to be done in the colony, have the necessary equipment to hand and keep working areas clear of unnecessary clutter. Always inform someone where you have gone. Keep your mobile phone handy and keep other people away while you are working with the bees.

Apiary Inspections

It is best to work from the farthest end of apiary towards the entrance. This way you do not have to keep going near colonies that have been disturbed, and they will settle more quickly. When working a beehive, you should stand behind the hive so that the foraging bees can come and go freely about their normal business.

Opening the Colony

It is not very easy to describe this in words. It is essential to see it in practice and to try it under proper supervision.

However, here are some pointers to help with that all-important first inspection. Some people like to smoke the entrance. I rarely bother these days but if you wish, give the entrance a few puffs of smoke and wait a minute or so to calm the bees (and yourself). When you are ready, remove the hive roof and set it upside down in a convenient spot within easy reach of your working area but not in your way. You may need to put the supers diagonally across the roof to avoid putting them directly on the ground.

Lifting the Boxes

If there are supers on the hive, insert your hive tool between the boxes, below those you want to remove, and lever upwards firmly. Before you lift the super boxes off, you must do two things. Firstly, lift the edge of the super up a little and smoke well into the gap. This is the point at which the bees will react to your entry into the hive because they see the crack of light and move towards it. A good puff of smoke at this point will calm them down. Then, while holding up the edge of the box, run the hive tool along to check that none of the frames from the lower box are stuck to the one you are about to remove.

Removing the Boxes

Only then are you ready to remove the box and put it onto the upturned roof. Give the bees a little more smoke to keep them below the tops of the frames. Just drift it gently across the top bars and in the places where you want to work. Some people like to use the cover cloths now to cover the top of the colony and keep things dark. However, they are not essential and the bees soon settle down once they are used to the light.

111

Removing Frames

Removing the first frame is always the trickiest bit because the frames fit quite closely together and there may be little room for manoeuvre. Always start at the edge of the hive closest to you. Sometimes it is easiest to start with the second frame in. There is not much spare room and there will be bees in the spaces so try not to roll them around.

Removing the first frame

▶ Loosen the first frame gently by levering it upwards a little under the lug to break the propolis seal

▶ Repeat this on the other side

▶ Once the frame is loose you can lift it carefully upwards to remove it from the hive

▶ Lift the frame slowly and smoothly, trying to avoid crushing or rolling the bees

▶ The first frame(s) usually contains only honey, or may even be empty, so the queen is unlikely to be on it

▶ Put the frame to one side in a safe place; I usually prop mine in front of the hive entrance

▶ This creates a gap that allows the next frame to be gently levered backwards and lifted out easily

Inspecting the Colony

Once you have removed the first frame, examine the rest systematically.

▶ Inspect the second frame, then put it back into the place from which you took the first frame

▶ Inspect each frame in turn, replacing it close to the previous one so you retain the space ready to lift out the next frame

- ▶ The working gap moves along with you as you continue your inspection
- ▶ Apply smoke gently as required to move the bees away from the frame lugs and to keep them inside the hive
- ▶ When the inspection is finished, move everything back into its original position and replace the first frame into the gap

Clear Purpose

The point of looking into a hive is to find out what is going on, so you should only open the colony with a clear purpose in mind. Being observant is an essential skill in beekeeping stockmanship.

Shaking Frames

There will be occasions when you need to remove the bees from the frame so you can see the comb properly. Hold the frame by the lugs with most of it in the gap and give it a sharp shake. This will dislodge most of the bees into the hive.

Recognizing the Components

The first thing is to learn to recognize the important components of the colony:

Shaking a frame

honey stores, pollen, sealed and unsealed brood, eggs, queen cells and the size differences between worker and drone brood. The colony is spherical in its organization, with honey stores towards the outer edges of the colony, a shell of pollen surrounding the brood and the developing larvae safely in the centre of everything where the heat is most easily controlled and the queen is safe. The different colours of pollen in the cells will reflect the variety of flowers your bees are working. Under normal circumstances, seeing eggs tells you the queen is present so it is not necessary to see her each time you inspect, although it is always reassuring.

Reading the Colony

It may take a while to understand the implications of what you are seeing. This is known as 'reading' the bees or the colony. The story of the colony is written in the combs for you to see once you know what to look for.

What to Look For

▶ Is the queen present and is she laying eggs normally?

▶ Is the colony developing as expected, and similarly to other colonies in the apiary?

▶ Does the queen have sufficient space to lay more eggs?

▶ Does the colony have sufficient room for honey storage?

▶ Does it have enough stores to last until the next inspection?

▶ Are there occupied queen cells present?

▶ Are there any signs of brood disease?

▶ What is the colony's Varroa status?

114

Keeping Records

The best memory in the world fails sometimes so it is helpful to get into the habit of keeping records. It may not seem important at the start, maybe with only one hive, but as the apiary develops, records become essential. You may want to look back at a particular colony's performance or compare colonies at the end of the season in order to plan for improvements or breeding from particular queen. There is also a legal requirement to keep a record of any Varroa treatments that have been applied.

Top Tip

Always hold the frame being inspected over the colony in case the queen should be on it and accidentally drop.

Learning from Others

Take advantage of opportunities to watch other beekeepers handling bees. It is just as instructive to observe beekeepers who aren't good at this as those who are. Lack of perfection is also an important learning process.

A frame with brood (centre), pollen (right edge) and honey (top edge)

115

Catching a Swarm

Swarming is colony reproduction, and a swarm leaving a colony is looking for a suitable cavity in which to establish its new nest. After leaving the hive, the swarm forms a cluster while scout bees search out a new nest site. Eventually, they 'decide' on a new home and the swarm flies there to take up residence. If you are prepared to collect a swarm, you could get some free, or nearly free, bees.

A swarm collection kit

Swarm Calls

Collecting swarms can be great fun but also very frustrating. If you are called to a swarm, try to get there as soon as possible. Swarms have a nasty habit of taking off for their new home a few minutes before you arrive!

The Swarm Cluster

Beekeepers are generally called to swarms that are clustering and this is really the only time it is possible to collect them easily. Once they have taken up residence in their new location they are very difficult to remove. To collect a swarm you will need certain items.

What You Will Need

▶ A strong container, such as a sturdy cardboard box, skep or purpose-made container in which to collect the bees

▶ A piece of loose-weave material, such as an old sheet, and some string

▶ A pair of secateurs or loppers

▶ A water spray

▶ Your protective clothing and smoker

Skeps

Skeps are very useful for swarm collection as they are strong but relatively light, and the bees can cling comfortably to the rough inner surface. Swarms can be collected in almost any container but make sure any cardboard box is strong enough not to collapse under the weight of the swarm, which can be several kilos. The cloth is used to wrap the container when it is being moved, so the weave must be close enough to prevent the bees from escaping but loose enough to allow air into the container, and it must be secured very carefully.

Collecting a Swarm

To collect a swarm, you need to transfer the queen and as many bees as possible into your container. It is essential that the queen is included or the bees will just fly out and reform

Top Tip

Before answering a swarm call, check they are honeybees, not wasps or bumblebees.

Collecting a swarm

117

their cluster. However, as she is likely to be in the middle of the cluster, if you can get the majority of the bees into your box, you are very likely to include her.

Collecting the Classic Swarm

The classic swarm hangs neatly from a branch and is really easy to collect.

Did You Know?

A well-made skep can support the weight of a man standing on it.

- Spread out the cloth near the base of the tree, ready to receive your box
- Lift the box so that it surrounds the swarm, then give the branch a sharp blow to dislodge as many bees as possible into the container
- Lower the container onto the sheet and turn it over carefully
- Prop up one edge of the container to make an entrance

There will be lots of confused bees flying around looking for the swarm. As long as the queen is in the box, the workers will bend their tails down to expose their Nasonov glands and start sending out a pheromone message to attract the flying bees. You can easily see this behaviour and it is a helpful indication of the queen's location. You may need to shake the branch again or even smoke the area to disguise the smell of the bees. If possible, wait until evening when the bees have stopped flying before you remove the prop, wrap the container in the cloth and secure it to prevent any bees escaping.

Hiving a Swarm

It is best to hive a swarm in the evening as the bees have the night to settle. You can hive a swarm in two ways, either by making the bees run up a board sloping up to the entrance or shaking them directly into the brood box.

Running in a Swarm

Running the bees into your hive is a lovely way of
hiving a swarm. Of course you will have already
set up the empty hive in its desired position ready
to receive the swarm. Use a board at least as wide
as the hive and long enough to be propped up to
the entrance at a slope of about 30 degrees. Cover
the board with an old sheet or other cloth, which
should hang over the sides of the board to prevent
the bees finding their way under the board. Turn
the box over above the sheet and knock the bees
onto it, making sure they have all been shaken out.
The bees will naturally run upwards looking for a
dark cavity, so they will start walking up to the hive.

Running in a swarm

Shaking in a Swarm

Instead of tipping the bees onto the sheeted board, open the brood box and remove six or seven
frames from the centre. Knock the container sharply to dislodge the bees into the hive then gently
replace the frames. Once the bees start climbing onto the
frames, replace the crownboard and roof and put in the
entrance block.

Did You Know?

Bees produce a 'footprint'
pheromone that marks where
they have walked and attracts
other workers to the spot.

Quarantining

If a swarm of unknown origin is coming into an established
apiary, ideally it should be kept separate until the beekeeper is
satisfied there is no disease in it. See the Pests And Problems
section for more on this (pages 220–21).

119

Feeding Bees

Bees are incapable of feeling hungry. Honeybees are programmed to collect and store plant-based glucose and fructose. They will do this when the conditions are suitable and will only stop doing so when these conditions cease. In good weather conditions, bees can collect an astonishing quantity of stores. When the weather turns cold, they cannot collect nectar and may need help.

Bees collect nectar and pollen, which are their natural sources of food

Understanding Feeding

It is essential to understand about feeding bees. However generous you are in leaving honey for the colony rather than harvesting it, there will be occasions when bees will need feeding.

When to Feed

The most important feeding period is in the autumn, to prepare the bees for the winter. Any newly formed colony will probably need a helping hand, especially if the weather is poor. A new swarm, a nucleus, a colony that has swarmed, a divided colony, or a colony being prepared to rear queens – in fact any colony where the normal balance has been disturbed – may require the beekeeper's intervention.

Checking the Stores

It is important that you check the food status of the colony at all times. It is difficult to give exact instructions about feeding because there is no single precise answer. The food the bees collect will depend on the weather and the available flowers. A beekeeper must be sensitive to the seasons and the size and vigour of each colony. This will mean checking the food stores and then acting to make up any deficit. As a rule of thumb, in summer the bees should always have a minimum reserve of at least 4.5 kg (10 lb), roughly equivalent to two solidly filled brood frames of honey. To survive the winter, the colony will need at least 18 kg (40 lb) of stores, or eight solidly filled brood frames.

Hefting

I test to see if the bees have enough food by hefting the hive. This simply means lifting it up just an inch or two at the back to gauge the weight. My guide is that if I can lift the hive easily it needs feeding, and if I cannot lift it at all, it has plenty (*see also* pages 141–42).

Putting a contact feeder onto the
crownboard to feed syrup

Sugar

Sugar is given to bees as a supplementary food to make up what they
have not been able to collect, or to see them through the winter after
the honey has been harvested. If you remove the bees' honey stores
you are responsible for making sure they have sufficient food to last
through the winter. The beekeeper should only take what is surplus to
the bees' requirements.

Feeding Bees Sugar Syrup

While honey is undoubtedly the best food for bees, sugar syrup is
a good substitute. Some beekeepers get very precious about the
correct syrup concentration. Many books tell you to feed sugar syrup
at different strengths, offering weaker syrup in the summer and more
concentrated syrup in the winter.

A Sugar Syrup Recipe

Personally, I have always found this far too complicated and for many
years I have used just one concentration with no apparent ill effects
on the bees. This is the standard '2 lb to a pint', that is, 2 lb of sugar to
one pint of water. Irritatingly, in this metric age, this does not actually
convert neatly into 1 kg of sugar to one litre, although increasingly this
is being offered as the recommended strength of sugar solution.

Safe Storage

The important thing is that the bees are able to reduce the water content to the required concentration
so that the sugar can be stored safely. They also add enzymes to invert it into glucose and fructose in a

122

similar way as for nectar. Some remarkable chemistry happens to glucose when it is dissolved in fructose at hive temperatures, making it perfectly formulated to store as a supersaturated solution for long periods.

Refined Sugar

The sugar used to make sugar syrup for bees *must* be white, refined cane or beet sugar. The bees do not have the enzymes to invert or digest more complex sugars or unrefined (brown) sugar. Feeding these substances will upset their digestive systems and make them ill with dysentery, possibly resulting in the loss of the colony.

Inverted Sugar Syrup

It is possible to buy ready-made inverted sugar syrups at the correct strengths and they are becoming increasingly popular. Weigh up the cost of obtaining inverted sugar syrup against that of sugar and the work and mess involved in producing sugar syrup.

Feeders

Feeders come in different shapes and sizes but essentially consist of a container for sugar syrup that allows the bees to feed from the syrup but prevents them drowning in it. There are four basic types of feeders.

- ▶ Contact feeders
- ▶ Rapid feeders
- ▶ Miller and Ashforth feeders
- ▶ Frame feeders

123

Contact Feeders

Contact feeders deliver the syrup in a controlled manner. They
usually take the form of a plastic bucket with metal mesh covering
a hole cut into a tight-fitting lid. This is filled with syrup and turned
upside down over the feed hole of the crownboard. Sometimes people
modify bottles or other everyday containers to work in the same way
as purpose-made contact feeders. The system works because the
air trapped above the syrup when the container is turned over forms
a partial vacuum. Some of the syrup will come out before the vacuum
forms, so first upturn the feeder over a bucket to catch the drips. Prevent
unwanted guests from gaining access to the feeder or the hive by placing an empty
super around it to keep everything bee-tight.

A contact feeder

Rapid Feeders

Personally, I do not have much time for rapid feeders. They
seem very fiddly and have bits that drop off too easily for
my liking. However, if you use one, it is placed over the
feed hole in the crownboard in the same way as a contact
feeder. Rapid feeders also need an appropriate surround
such as a super to keep out robbers.

Ashforth and Miller Feeders

These feeders follow the same basic principle as the rapid
feeders. The main difference between the Ashforth and
the Miller is the position of the access slot, at the side
(Ashforth) or in the centre (Miller). They are designed to fit
neatly across the whole hive, doing away with the need for

A rapid feeder

extra supers to keep out robbers. This is a great bonus if you have an out apiary. On the downside, the bees can often be slow to use them, especially if the weather is cool. The hive must be absolutely level and the wooden types can be prone to leaking. Be very sure they are 'bee-tight' before using them.

Frame Feeders

A frame feeder usually replaces a brood frame and is essentially a wooden box fitted into a frame. The box contains a float for the bees to stand on so that they can lap up the syrup without drowning. They only deliver a small quantity of food and require entry to the hive for refilling so their use is usually limited to specialist procedures.

A Miller feeder

Do's and Don't's

It is important to keep the sugar syrup inside each hive. This will protect the bees from the dangers of robbing by other bees and will also protect from marauding ants. Generally, syrup is fed in a special feeder over the feed hole in the crownboard and inside an empty super.

Evening Feeding

It is important to feed in the evening to minimize disturbance in the apiary and, ideally, every colony should be fed at the same time. Do not spill any syrup as this will alert the bees

Top Tip

If in doubt, feed your bees. You cannot feed them too much (because they will just store it for later) but it is certainly possible for them to have too little food.

and set them searching for the food. Remember the weather needs to be warm enough for the bees to be able to reach and process the food. They will not take food into the hive in very cold weather, although contact feeders make this process easier. Leave feeders in place until they are empty.

Feeding in Winter

If the bees were well fed in the autumn, they should not need feeding during the winter but, for whatever reason, this is sometimes necessary. Traditionally, solid sugar candy was fed. This is made by boiling a sugar solution (without caramelizing it) until it is thick enough to pour into moulds and set. An easier alternative is to find a baker who sells baker's fondant. Candy can also be purchased from bee equipment suppliers.

Feeding Candy or Fondant

Candy is solid so it is just cut into chunks and put into any suitable container (such as a plastic ice cream container), which is placed upside down over the feedhole in the crown board. This is so the bees have direct access to it from within the hive. Extra space will need to be created under the roof to accommodate the container of candy. This is done with an empty super box under the lid in the same way as for contact feeders.

Top Tip
Keep sugar feeds inside the hive: it is easier to prevent robbing than to stop it once it has started.

Checklist

▶ **Choose and plan your site carefully:** Apiary sites must have sufficient room for equipment and colony manipulations. Consider other people (i.e. neighbours) when setting up an apiary.

▶ **Give access to water:** The bees must be provided with a suitable water source.

▶ **Provide shelter:** Hives need to be protected from the weather.

▶ **Secure your hives:** Hives should be put on stands that are sturdy and stable.

▶ **Plan your movements:** Bees should be moved in the evening and less than three feet or over three miles.

▶ **Handle your bees gently:** This is important for both you and your bees' health.

▶ **Assembling your hives:** The wisest choice is to assemble your hives from bought flat packs – everything will be the correct dimensions, whilst not being as expensive as buying them ready-assembled.

▶ **Frames:** Put bought wired foundation into your frames. Keep spare frames and foundation.

▶ **Know your equipment:** Practise lighting the smoker until you can do it easily and use the right fuel (choose a smoker fuel that gives a cool smoke).

▶ **Inspecting the colony:** Only open a colony if you have a good reason and learn to 'read' it well. Keep records – this will help you learn and plan improvements.

▶ **When collecting a swarm:** Make sure the queen transfers to the container.

▶ **Monitor food stores:** Be aware of the level of food stores in your colonies and feed sugar syrup or candy when necessary. Don't encourage robbers by spilling sugar syrup in the apiary and make sure hives are bee-tight.

127

Through the Seasons

The Beekeeping Year

We will now take a short stroll through the beekeeper's year. However, bear in mind this calendar will vary depending on the geographical location, local weather patterns and the types of forage available for the bees. Beekeeping activity is utterly dependent on weather conditions, so the following is necessarily only a guide. It is one of the more alarming facts of beekeeping that winter starts in early autumn. The harvest is in and the bees need treating for Varroa mites and feeding to settle them for the winter.

Take care when harvesting honey to make sure to leave enough for the bees during winter

130

Autumn

Feeding the bees is likely to be uppermost in many beekeepers' minds. As with everything else in beekeeping, it is really difficult to make generalizations, because standard information has to be adjusted to fit your own management scheme and local environmental and weather conditions.

Winter Stores

We have already seen that a normal colony of bees will need at least 19 kg (40 lb) of stores to get it through the winter. Some of this may already be present in the colony as honey in the brood box. Sometimes a strong colony of bees will have used all the brood area for breeding and stored all the honey in the supers (and perhaps these have been extracted by the beekeeper), so take care when harvesting honey to ensure the bees are left with enough food for their needs. The bees must come first.

Supers and Supplements

Some beekeepers like to leave a super of honey for the bees and others collect a heather crop specifically as winter feed. Remember, you can heft the colony (*see* pages 141–42) or even weigh it with a spring balance on each side if you are unsure whether it has sufficient stores.

Top Tip
Make sure the bees have sufficient food left after the honey harvest, and never harvest honey from the brood box.

Time to Feed

When you feed is also important. If you feed too early, you run the risk of stimulating the queen to start laying eggs and creating an extra demand for the winter supplies, possibly causing them not to last the season. If you leave it too late, the weather may have turned too cold for the bees to ripen the stores properly, thereby running the risk of dysentery and the spread of Nosema during the winter.

Fuel and Insulation

Plans can be ruined by a long 'Indian summer' and you may need to check there are still sufficient stores if this happens. A colony with ample stores is not only set up for the winter with energy-rich food, but the stores also act as excellent insulation, so a well-fed colony is doubly protected against the rigours of winter.

Supersedure

This is the time of year when you may see a supersedure queen cell. If you come across a late queen cell, close the hive down carefully and leave things well alone until spring. (*See* pages 52 and 238 for how to recognize a queen cell). There are two types of supersedure; these have been termed perfect and imperfect.

▶ **Perfect supersedure:** Perfect supersedure is where the old queen continues to lay eggs until after the new queen has hatched and mated. She then gradually fades away, leaving the way clear for the new queen to take over the colony. This is one of the circumstances when two queens may be seen together in a colony.

▶ Imperfect supersedure: Imperfect supersedure is a much more dangerous strategy for a colony. This is where the old queen dies before the new queen is mated and laying, leaving the colony at great risk if anything

A queen bee is clearly visible here, marked with red paint

should happen to the new queen or she is poorly mated. This is frequently why colonies are found to be queenless in the spring, and another reason why it is a good idea to keep at least two colonies so things can be more easily rectified.

Late Autumn

By now, only the last remnants of the beekeeping season remain. The 'season of mists and mellow fruitfulness' is passing, the harvest is safely collected and the bees are tucked in for the winter. It is essential to make sure the bees are well provisioned and autumn Varroa treatments are completed. Sometimes there is an 'Indian summer' and the bees continue to collect from the last flowers.

Inserting a Varroacide strip – Varroa treatments must be complete by late autumn

Late Flowering Ivy

I am always very pleased if the bees are able to collect a late crop of ivy nectar. The sugars in this source are highly concentrated and attract swarms of insects to the flowers. The honey does have a reputation for granulating in the bees' stomachs, causing their uncomfortable demise. However, my own experience leads me to consider this an old wives' tale. Indeed, I encourage the ivy on the trees close to the apiary.

Collecting pollen and nectar from ivy

Ivy Honey

For many years now, I have been very glad of the ivy as a source of winter stores, often gaining as much as 25 lb of honey in addition to the sugar I have fed. I have rarely lost a colony over the winter except from my own carelessness and I am certain that I have never lost one because of problems with ivy honey. It seems to me that the bees have been around much longer than we have and maybe they know best what suits them.

Testing for Acarine and Nosema

If beekeepers have been careful, they will have checked their bees (or had them checked) for Acarine (caused by Acarine, or 'tracheal', mites) and Nosema. Both of these diseases are very debilitating and will only get worse during the winter and early spring when the bees are confined for long periods in close proximity to each other. It is not too late to test and there is still time for treatment. Local associations sometimes have 'microscope days' to do this.

Let Colonies Die Out

My own feeling about Acarine is that it is better to let bees that are susceptible to this condition die out, as

Dead bees – not what you want to find in your hive!

there is a strong genetic resistance to it which is best developed by breeding from resistant colonies and despatching susceptible ones.

Nosema and Dysentery

Nosema can be treated if necessary. Dysentery will exacerbate the spread of Nosema spores but is not a cause of Nosema (or vice versa). Nosema is caused by a microsporidian fungus that damages the bees mid-gut, whereas dysentery is the result of incorrect feeding which upsets the bee's digestive system. Avoid feed that is too weak or given too late, or that will ferment or not be ripened adequately for winter storage, as this will cause dysentery. Heather honey or honeydew (*see* pages 178 and 182) left on for winter feed are also often blamed for dysenteric problems. Inappropriate or impure feed, such as brown sugar, will also cause dysentery.

Looking for the queen

Ejecting the Drones

It may still be possible in some places to see drones being tipped out of the hive, having spent their summer loafing round the hive and chatting up queens. Although it is quite possible for drones to be present in hives very late into a mild winter, it can also be a sign that all is not well in the hive, even that the queen has failed. An early inspection in spring will be needed if you suspect that this is the case.

135

Winter

For many people this is the time when they can look back over the year and enjoy the fruits of their labour. All that is left is to make those candles or test the mead. Uncapping honey for harvesting (*see* page 170) does leave you with a good crop of wax, so get on with that candle-making for Christmas.

Winter Checks

The bees will now be dormant for the winter and the beekeeper's main task is to make periodic checks, especially in the case of out apiaries. Prompt action can save a colony from being lost. Long-lasting snow can sometimes cause problems. A colony can be suffocated by debris, dead bees and snow blocking the entrance. Don't forget, there can be vandals (both the two-legged and the four-legged kind). For the first time ever, we have started getting badgers in our garden, so I am being extra-vigilant as they can do great damage to beehives.

Winter Varroa Treatment

One key procedure at this time of year is to administer a Varroa treatment of oxalic acid. This is used to complement autumn treatment with thymol. Oxalic acid is toxic to the brood so needs to be applied when none is present. Oxalic acid is normally sold ready mixed as a 3.2% solution in sugar syrup. Fill a 50-cc syringe with warmed oxalic acid. Take the crown board off the colony and trickle 5 cc of oxalic acid down each seam of bees. It is not an exact science, so give a little more to the fullest seams in the middle and little less to the smaller seams at the edge of the cluster. The whole procedure takes less than a minute per hive, but it is best to wait a while if conditions are subzero. A temperature of 8–10°C (46–50°F) is ideal.

Administering oxalic acid treatment – ideal to treat while bees are in the winter cluster

Seasonal Supplies

One very important task a beekeeper can do over the winter is to make regular 'quality control' checks on the supplies of mead, honey beer or sloe gin. Do not forget we are entering the Christmas season of goodwill and a pot of honey will help to keep the neighbours sweet, while a beekeeping book in the Christmas stocking will go a long way toward ensuring the wellbeing of your bees next year.

137

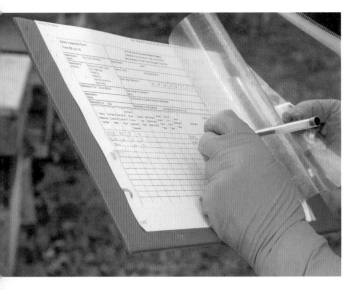

Record Review

If you are feeling studious, now is a good time to review your beekeeping records. This is particularly important if you are making an assessment of your bees to plan your queen-rearing next season. Even a simple analysis of temper and productivity will be valuable in identifying stocks that you want to breed from, to requeen or to cull.

Selecting Breeding Stocks

Beekeepers should be choosy about the numbers and quality of the stocks that they overwinter. In these days of extreme weather events, it is important to know which bees are tough and thrifty enough to survive. Colonies that can produce a crop in a cool, wet summer and ones that survive in mild winters are particularly precious. Bees that perform in a 'challenging' year deserve your attention as breeding material next year. They may not always give the most honey when conditions are at their best but, over time, these are the ones that will give reliable and consistent crops for the smallest cost.

Late Winter

By now the bees are normally clustered comfortably together. They are not actually hibernating. There is a little miracle happening here as they generate their optimum winter temperature. Despite the fact that bees are cold-blooded creatures, they can still maintain very exact thermoregulation in the colony over a wide ambient temperature range.

Generating Heat

Within the winter cluster, the bees in contact with the storage combs consume honey and then utilize this energy to generate heat. This is achieved by a curious isometric 'vibration' of the huge flight muscles that almost fill the bee's thorax. The whole of the cluster expands or contracts depending on whether it needs to be warmer or cooler. The outer bees act as insulators to the inner bees, and the inner bees act as heaters for the outer bees.

Keeping in Touch with the Stores

The cluster has to loosen occasionally to move to another area of stores. If the weather is too cold to allow this, the colony may die of starvation, even in the midst of plentiful food supplies. The beekeeper doesn't really need to worry about all this because the work to ensure that wintering goes smoothly should have been done months ago in the autumn. Bees should be flying on a sunny day when it is warm enough. They will defecate and may be able to collect pollen from winter-flowering plants.

Bees may be able to collect pollen from some plants in winter and early spring

Minimal Beekeeping

In my opinion, the extent of beekeeping activity at this time of year shouldn't go beyond a stroll round the apiary to work off the Christmas excesses and to make sure the hives are safe. Are they properly mouse-proofed? Is snow or debris blocking the entrances? Have they been knocked over by wind or

139

animals? One person I knew had their hives washed away in a flood one winter, while another had a hive completely blown away. Remarkably, the latter beekeeper put it back together (albeit with the brood box upside down) and the bees all survived.

Time to Study

This is a good time for beekeepers to relax with a good bee book and a glass of mead. Take advantage of the time to read, study, ask questions, attend lectures or whatever it takes to learn more about the bees and plan what is to be done next season.

Moving Hives

If you want to move your hives within the immediate vicinity, this should be done during the coldest months. During the winter the 'less than three feet or more than three miles' rule can be ignored for relocating bees. Choose a long frosty period so that the bees come out more slowly and carefully and re-orientate gradually when they start to fly again.

Spring

When I first started learning about beekeeping, one of the things impressed on me was that the bees' survival is at a critical point at the start of spring. The first stirrings within the hive mean that the colony is balanced on a knife's edge and adequate food supply is the key factor. This is where it becomes apparent whether your autumn preparations were sufficient or not.

Brood Rearing Commences

Within the cluster, things are changing. The first cells are being prepared for eggs and the queen is coming back into lay. As brood is produced, the nest temperature rises and this, in turn, produces more brood-rearing activity. The extra brood-rearing stimulates the demand for food, and the energy and protein requirements of the colony soar. Until now the colony has been 'ticking over', needing no more than 2 kg (5 lb) of honey a month. Once brood-rearing really gets underway, the demand will be more like 2 kg (5 lb) a week. Now is the time when ample autumn provisions really count.

Hefting a hive should give an indication of whether enough food stores remain (*see* overleaf)

141

Hefting the Hive

To check whether enough stores remain, hives can be hefted. This involves the estimation of hive weight by lifting the box at one side, just off its stand. Most books suggest you lift first one side of the box and then the other, and then add the weights together to give an idea of the quantity of stores remaining. It can be done using spring balance scales if you really must. However, this seems unnecessarily complicated to me.

A Guide to Hefting

My own method is to lift up the back of the hive. If I can tip the beehive easily then there isn't enough food; if I find it hard to lift then there will probably be enough. Knowing where the balance lies comes with experience, so it is a good idea to heft hives at intervals to get a feel for it. However, discovering there is not enough food at this point is a bit like finding the horse has gone when you open the stable door.

Emergency feeding using a bottle contact feeder

Emergency Feeding

The best way to prevent this miserable state of affairs was to feed well last autumn. However, there are times when things go dreadfully wrong and it can be helpful to know how to retrieve the situation. If you really feel that your bees will die if they are not fed, you must do so. A feed of warm sugar syrup could do the trick. This has to be done using a contact feeder where the sugar is available as a 'drip' for the bees. It also has to be warm and, even then, the bees may ignore it. Candy may also help in these circumstances.

Feeding Candy

All candy needs a lot of water to render it useful to the bees. This requirement for water and the energy candy supplies stimulate activity when, in fact, the bees should be quietly clustered. It wakes them up, but in the wrong way. In effect, it is a bit like being brought a cup of tea at one o'clock in the morning.

Woodpecker Protection

In some areas the green woodpecker can present problems. At this time of year the supply of insects is at its lowest point and a beehive must seem like woodpecker heaven, making a tasty and easy food source if not properly protected. It is only the green woodpecker that does this. The little black, white and red jobs are just fine. If you are likely to have problems with woodpeckers, slip hive protectors made of small-mesh chicken wire over the hives. Wrapping hives in plastic will also work, but may cause problems with condensation.

Using wire mesh to protect from woodpeckers

Late Spring

The beekeeping year really starts to get underway as spring develops and both bees and beekeepers are stirring. The first generations of fluffy new bees will have hatched and be taking their first tentative flights, re-establishing the colony's cycle of life. The amount of brood is increasing dramatically and the protein in pollen becomes particularly important to the colony in order to ensure the growth of the new larvae.

Providing Pollen

There are very few areas where there will not be enough fresh pollen available for the colony. Some beekeepers like to provide pollen 'patties'. An increasing number of useful bee plants are now in

flower but the most important by far at this time of year is the willow, especially in rural areas. Willows of all types provide both nectar and pollen in abundance, giving a welcome boost to the bees' spring development.

The First Inspection

Do not look at the brood nest too early or the bees may kill the queen. As long as you can see plenty of pollen going in at the entrance and the hives are a good weight, you can be sure that the queen is laying and things are okay. Depending on where you live, March inspections may not be entirely necessary, especially if the weather is cool. Wait for a warm day.

Water Availability

Of prime importance is a water source. There is a clear relationship between brood rearing and the water requirements

Bees drinking

of the colony. Raising brood demands lots of energy. At this time of year, this energy comes mainly from stored honey that must be diluted before it is used. A clean water supply close to the hive will be invaluable to the bees. It will also help to train them to the water source you provide now – rather than your neighbour's pond. Remember, provide a platform so that the bees do not drown.

Spring Dwindling

Another problem that may arise at this time of year is spring dwindling. This is usually blamed on Nosema, but pigmy shrews and great tits can also cause problems that are largely unreported and unrecognized.

Bees drinking from floats in a trough

Top Tip

As soon as flowering currant (*Ribes sanguinium*), often called the beekeeper's barometer, is in flower, you should be able to inspect your bees.

The Start of the Active Season

Spring heralds the start of the active beekeeping season and the bees will need your undivided attention from now on. The time is right for a major spring inspection to see what problems might have arisen over the winter and to give the hive a good spring clean.

Varroa Monitoring

Traditionally, spring is the time to change hive floors. These days most beekeepers have open-mesh floors for Varroa monitoring

145

which need careful cleaning to prevent wax moth building up. Using monitoring floors to establish Varroa infestation levels will help you plan this year's attack on the wretched mite. Do not be afraid to talk to your local bee inspector or the local association for the latest ideas for Varroa control. Monitoring methods and an Integrated Pest Management Approach (IPM) to Varroa control, essential with the advent of pyrethroid-resistant mites, are detailed in the Pests And Problems chapter (pages 201–05).

An open-mesh Varroa floor

Spring Brood Check

It is really important to conduct a thorough spring brood check. Good health and a good queen are vital to the bees at any time of year. Drone-laying queens or queenless colonies will be obvious, although at this time of year it is not so easy to replace a failed queen unless you want to make do with an imported one. At the other extreme, where colonies are really strong, it is quite possible to have a very early swarm. However, this is not a disaster as the bees will have a vigorous young queen and there is less likelihood of the colony swarming again.

Forager Mortalities

It is not uncommon for foraging bees to be caught out by the vagaries of the weather during early spring. A sudden rapid temperature drop can leave them unable to get back to the hive, leaving an obvious scattering of dead bees. In many places, an important forage crop at this time is the dandelion with its characteristic golden-coloured honey and wax and bright pollen.

Oilseed Rape

Oilseed rape begins to bloom now and these flowers, in some form or another, will be available for much of the spring and early summer. This bright yellow-flowered field crop is really important to lots of beekeepers, and in many places the mile upon mile of this crop is the most important nectar source of the year. However, the large quantities of honey it yields can granulate quickly.

Late Spring

Brood rearing is now in full swing. Plentiful forage means the demands for pollen and nectar are easily met as long as the weather is good. In many areas beekeepers expect to remove a honey surplus from the spring flow. Good weather now makes it an excellent time to ensure foundation is drawn out.

Providing Sufficient Space

Space is a vital factor at this time of year and lack of it can provoke swarming. There must be plenty of space to keep up with the queen's massive egg-laying effort, and extra space in the form of supers should be given well ahead of requirements.

Adding supers

Adding Supers

When you see the bees have spread out to the outer edges of the super frames, it is time to add another super. There is always lots of discussion about whether the new super should go on top of the old one or underneath.

Under or Over?

Traditional wisdom has it that the bees will go into the new super more quickly if it is put on underneath the full one, and this will be beneficial to the bees because they exploit the space much more rapidly. The supporters of the other camp argue that the bees go into them quickly enough at this time of year, and to put the supers on the top is easier work as a quick glance will show whether you need another super, because you only need to lift the crownboard.

Regular Inspections

Regular inspection at this time of year is essential. The swarming season is well under way. Only large, well-provided colonies swarm. Queen rearing can begin once the worker population has reached about 12,000, with swarming occurring once 20,000 workers are present or the bees are occupying about 40 litres of hive capacity.

Watch for Swarms

The loss of swarms severely depresses honey production, as a significant portion of the foraging population of the colony leaves to form the new one. Remember, 'A swarm of bees in May is

Did You Know?

The nectar flow from oilseed rape is a wonderful stimulus for the bees to draw out foundation.

148

worth a load of hay' while 'a swarm of bee in July, ain't worth a pig's eye'. To catch one early is to gain something very valuable from someone else's carelessness. However, to lose one means you may have compromised your own honey harvest.

Swarm Control

Your winter planning should have included a decision on which swarm control method you are going to adopt and you should have ensured you have enough equipment to hand. There is nothing worse than trying to put frames together in a rush while the swarm fills the cardboard box with comb, or dashing miles to the local equipment supplier for more foundation, only to find they have run out.

Sealed queen cells indicate that the swarm has gone

Different Methods

All the 'beekeeping greats' have invented a swarm control method. Essentially, however, all swarm control methods divide the brood from the flying bees, thus mimicking what happens naturally. With all methods, it is really important to remember that if sealed queen cells are present, the swarm has gone and, with it, the old queen. Thus, if you destroy all the queen cells you are left with a queenless colony. This is great if you wanted a queenless colony but a nuisance otherwise!

Did You Know?
The old queen leaves with the prime swarm as soon as the first new queen cell is sealed.

149

A Simple Swarm-control Method

▶ Wait until you see unsealed queen cells (*see* pages 52 and 238 for a description of queen cells)

▶ Move the parent hive to a new permanent stand at least four feet away

▶ Place a new brood box (or nucleus box) on the original stand

▶ Select a comb with a good-sized, unsealed queen cell

▶ Gently brush every single bee from this comb and destroy all the other queen cells

▶ Put the selected comb into the new brood box; the flying bees will find their way back to this box but the old queen cannot be present as no bees were transferred

▶ Add at least two frames of food stores (and pollen) plus one frame of sealed brood – brushing off all the bees first

▶ Fill up the new hive with frames of drawn comb or foundation and reduce the entrance

▶ Replace the frames removed from the old box with frames of foundation; this will both improve the ventilation of the hive and give the remaining bees something to do apart from thinking about swarming, although the reduction in population should quell the swarming urge

▶ Divide the supers between the two hives

▶ Feed both parts as necessary

It will take about three weeks before the new colony has a functioning queen and it is very vulnerable during this time. Keep a careful eye on things without disturbing it too much.

Carrying out a shook swarm: transfer of
bees to the new brood box

Summer

This is the peak of the season for beekeeping work. It is very difficult to make generalizations about beekeeping because it cannot be dissociated from the local environmental, climatic and agricultural conditions. This is particularly relevant in this season. Your beekeeping agenda will be in part determined by the agricultural practices in the area where you live. We will start by looking at two broad pictures: one for beekeepers who live in areas where oilseed rape is the major nectar source, and one for those who live where it is not important.

Oilseed Rape

Oilseed rape can produce enormous crops of fine-textured, very white, mild-flavoured honey. However, the major problem is that its high proportion of glucose to fructose means that it granulates very quickly and sets into what I term 'Geller honey' (that is, it can bend spoons). It will quickly set in the comb.

Dealing with Oilseed Rape Honey

If the beekeeper faced with this problem wishes to remove it by extraction rather than sacrifice the drawn comb, the honey must be extracted as soon as it is 'ripe'. Ripeness of unsealed honey can be checked by gently shaking a comb. If no honey flies out, it can be extracted. As soon as the hive starts to cool (because of a swarm leaving perhaps), the honey will start to granulate. As a rule of thumb, rape honey will granulate within 10 days of being sealed in the comb, so it is really important to harvest it as soon as possible.

Urban Areas

For beekeepers who live in more mixed agricultural areas or in towns, the season is rather different. The main honey flow has not yet started. Town bees usually do very well as the variety of forage sources found in town gardens ensures a constant and reliable nectar flow. Urban temperatures are often higher and the microclimate more amenable than rural areas.

The 'June' Gap

Occasionally, between early and late honey crops, there can be a gap in the available forage known as the 'June gap'. It is not a regular annual beekeeping feature, but it is worth keeping the proverbial weather eye open just to make sure the bees are bringing in enough forage to keep the colony thriving at this time of year. Beekeepers with the yellow Italian-type races of bees will really notice a June gap if the weather deteriorates, since these bees are renowned for their inability to adjust their brood production to prevailing environmental conditions. Such beekeepers should keep a close eye on their colonies' stores.

Swarm Inspections

The swarming season is not yet over, and the weekly inspections for swarm control are still vital. This is especially so if your main honey flow is expected in July, since it is now that your future foraging force is developing. However, if the bees have already swarmed or you have successfully controlled swarming, the need for weekly inspections is coming to an end. The colony is unlikely to swarm twice in one year. If it makes more swarm preparations, you should definitely look at replacing the queen.

Top Tip

Many wildflowers are important honey sources – so don't think of them as 'just weeds'.

The Main Honey Flow

In more traditional agricultural areas and most urban areas, there is a defined but short period in summer where most of the surplus honey is collected. Beekeepers call this the main honey flow. All the hard work put in to prevent swarming, and ensure that brood rearing is timed so that the foraging population will coincide with the peak flowering time, pays off now. You do not need to add more supers at this time of year unless you know that your bees need them. From now on the brood nest is beginning to contract.

Summer Forage

Important summer plants are the wild plants and trees such as bramble, lime, sweet chestnut, clover and rose bay willow herb. The mixed floral honey they produce is particularly flavoursome. These plants often predominate in

153

areas where the agriculture is based on livestock rather than arable production and is also a good reason why beekeepers should not keep things too tidy.

Clover

Clover can be a picky sort of thing where nectar production is concerned, needing a temperature of at least 20°C (70°F) to flow and adequate moisture in the soil if the nectar isn't to dry up too quickly. Drought conditions can affect the nectar flow in many plants even though prevailing temperatures may be perfect. Nothing is ever predictable in beekeeping. But when conditions are right you can hardly walk a step across a grassy field for fear of treading on a bee in clover.

Summer Honey

It is lovely to think of all that honey pouring into the hive and the best thing ever is to stand in the apiary after a gloriously warm summer's day and just listen to the bees fanning the water off the nectar and processing it into honey. This has got to be the ultimate sound of summer. Those brambly corners give a double benefit of honey and blackberries too – what a good deal! However, the way the nectar flows and the bees' ability to collect it depend absolutely on the weather. 'Not sunny, no honey' you could say.

Late Summer

This is the time to harvest a traditional summer wildflower-based honey crop. There is nothing nicer than a bumper crop of honey in beautifully sealed super combs as you stagger indoors under the weight of your bounty, even though the kitchen floor is sticky for weeks and you have put on pounds round the middle from eating great scoops of the honey and (medicinal of course, in case of hay fever!). The smell and the taste are never better than fresh from the hive. I set aside a dedicated time for honey harvesting each year, because honey is easiest to extract straight from the hive. Keep the extracting room warm too, as honey becomes more viscous as it gets cooler, which makes it slow to flow and hard to strain.

Leaving honey or comb fragments around the hive risks robbing

Robbing

Honey harvesting can bring problems with robbing. Not only does this create a danger to vulnerable colonies that can be completely wiped out but robbing is also an excellent way of spreading disease. Take great care not to allow robbing to start. Prevention is infinitely better than cure. Pay extra attention not to spill honey or comb around the apiary.

Top Tip

To avoid problems with robbing and disease, never allow bees any access to supers of honey at any point before, during or after harvesting.

Keep Tidy

Under no circumstances should wet supers or cappings be left in the open for the bees to clean up. This is just asking for the bees to rob each other. This can result in disease spreading between colonies or, at worst, the weakest colonies being completely destroyed by a stronger one.

Replacing Supers

If possible, 'wet supers' (that is, supers whose combs are very sticky with the remaining traces of honey after the honey crop has been extracted) are best dealt with by replacing them on the hive they came from, over the crownboard with the feed hole open, so the bees can take back the spare honey, leaving the super nice and dry ready for storage.

Testing for Queenlessness

Late summer is often the time when people panic that the queen may not be present. This is because her egg-laying rate may have reduced dramatically and there may be no signs of eggs. The black bee races and their hybrids are characterized by a brood break at this time of year. However, it is also possible that there is a problem, and now is the time to fix it. Do they need a new queen? It is a difficult judgement to make, even for an expert beekeeper. A simple test is to put in a frame that contains eggs from another colony. If they build queen cells then the queen was lost; if not, they will simply rear the brood as their own. This is one of the main reasons for keeping a minimum of two colonies.

Checking mite drop on an insert

Varroa Treatment

Removing honey at the end of the summer gives time to sort out autumn Varroa treatments before the weather closes in. More colonies are lost from treating too late than from any other cause (except not treating at all, of course). Remember, it is the brood that is being raised from now onwards that will comprise the winter bees needed to take the colony safely through the winter.

Checking Mite Levels

It is really important that the winter bees are not damaged by Varroa or viruses. In any case, long before the days of Varroa, I was always taught that the honey the bees collect before the summer's end was mine; after that it is theirs and an important contribution to their winter survival. See pages 203–05 for how to check Varroa mite levels.

Uniting Healthy Colonies

Large colonies will survive the winter better than small ones. Small colonies can be united with another colony as long as you are sure they are healthy. This is a very easy operation. Place a sheet of newspaper over the 'host' brood box and put the brood box that is being moved on top. This prevents the two colonies from fighting each other by allowing their odours to mix. Two sheets of tabloid or one of a broadsheet will be sufficient, depending on how high-brow your bees are. You can punch a few holes if you like, but the bees will chew the lot anyway.

Placing newspaper on the 'host' brood box prior to uniting two colonies

United Under One Queen

Once united, the frames from the two boxes need rearranging into a single box. Keep the best combs and remove the others. You do not need to single out a queen unless one is particularly bad-tempered and you want to remove her. Otherwise, the two queens will fight it out and, hopefully, the best one will remain. The books will say the young one will kill the old one, but I would not bank on the bees reading the books.

157

The Beekeeper's Year

Here is a quick summary of what your bees will be doing and how you should be looking after them through the year. Remember to adapt this to your own particular location and climatic conditions.

Winter (e.g. November–February in the UK)

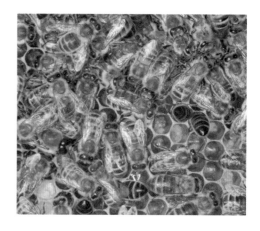

- Bees cluster in the hive
- Apply oxalic acid to kill Varroa mites during the brood-free period, usually around the end of December in countries such as the UK

Very Early Spring (February)

- Brood rearing starts
- Check stores by hefting the hive (and feed if necessary)

158

▶ Prepare supers, frames and foundation for summer use

Early Spring (March)

▶ Make sure all your colonies are flying well in suitable weather conditions
▶ On a warm day, use a little smoke and lift the crownboard.
The bees should look as strong now as they did last September

New frames of foundation

▶ Visible excreta on the top bars is a sign of dysentery
▶ Heft the hive and, if necessary, feed dilute syrup in a contact feeder
▶ In early districts, watch for signs of swarming from late March

Spring (April)

▶ First spring checks
▶ Clean the floors
▶ Keep a close check on the food supply
▶ Check if the queen is okay
▶ Check if there is enough space
▶ From mid-April, start regular brood nest inspections of larger colonies

159

▶ Add a queen excluder and super when the brood box is full of bees

▶ Remove old, broodless combs and replace with frames of foundation

▶ Feed the colony if necessary to help them draw out the comb

Late Spring–Early Summer (May–June)

▶ Make regular weekly inspections for swarm cells

▶ Regularly check the super space. Add supers if needed

▶ Prepare equipment for swarm control

▶ A spring honey harvest may be possible

▶ Check the bees are healthy

▶ Monitor Varroa status

▶ Use biotechnical Varroa control methods if required

Mid–Late Summer (July–August)

▶ Swarming should be over by mid–end July

▶ Harvest honey at the end of July/beginning of August

▶ Make sure hives are bee-tight

▶ Reduce entrances to prevent robbing

▶ Do not leave frames or honey around in the open

▶ Check Varroa status

▶ Start Varroa treatment if using Apiguard®, Apistan®, ApiLife Var®, Thymovar® or Bayvarol®

Early–Late Autumn (September–October)

▶ 16–19 kg (35–40 lb) of stores are needed for winter survival

▶ Feed bees sugar syrup (from the end of September until the colony is heavy or will not accept any more food)

▶ Make sure hives are bee-tight and wasp-tight. Keep entrances small to prevent robbing

▶ Are the colonies big enough to survive? They need 5–6 frames of brood to be safe

▶ Unite colonies if necessary

▶ Are the hive parts in good condition?

▶ Are the bees healthy?

▶ Remove Varroa treatments according to the manufacturer's instructions

▶ Record the batch numbers of treatments used and the dates of insertion and removal. This is a legal requirement

▶ With deep-floored hives, remove the entrance block and fit a mouse-guard by the end of October

Feeding using an Ashforth feeder

161

Reaping Your Rewards

About Honey

The honey harvest is the point at which all your efforts and care for the bees are rewarded by a share of their crop. Nothing tastes better than honey, from your own bees and local flowers when it's fresh from the hive. The double dividend is a crop of beeswax too. If you are really keen, you can also collect the medicinal propolis.

What Honey's Made Of

Honey's raw materials, the nectar and other natural plant exudations, have been collected and converted into a delicious food that reflects the plant sources it came from. Honey is essentially a supersaturated solution of glucose dissolved in fructose derived from the cell sap of the plants. Miraculously, at exactly hive temperature it is at its maximum solubility.

Typical Honey Composition

▶ 18 per cent water
▶ 35 per cent glucose
▶ 40 per cent fructose
▶ 4 per cent other sugars
▶ 3 per cent other substances

Did You Know?

The ideal temperature for storage of honey is below 10°C (50°F).

Minor Components

What gives honey its unique properties are the 181 minor components so far discovered, some of which are unknown anywhere else. Its exact composition depends on its plant sources and no two honeys are identical. Dark honeys usually have the strongest flavour while milder-tasting white honeys often command a higher price. This is partly because honey darkens over time or with heating (so dark honey could be old or damaged) and partly reflects traditional preferences for light-coloured honeys. However, heather honey, which is a rich, port-wine colour, has to be one of the best that is ever harvested.

Medicinal Properties

Honey has significant medicinal properties and uses. It is used with traditional and herbal medicines and there is increasing interest in using it in conventional medicine. Honey is valuable for burns and wounds and helps skin heal, especially in leg ulcers and bedsores, because of its antibiotic and debriding effect. It can be used to alleviate conjunctivitis: two drops will dissolve in the fluid of the eye and act as an antibiotic. As a soothing honey and lemon or honey and whisky drink, it also relieves sore throats, coughs and colds.

Keeping Honey

Honey can be kept almost indefinitely as long as it is harvested correctly and stored carefully. It does not 'go bad' like perishable foods but the subtle flavours that characterize fresh honey are lost over time, and heating honey hastens this process. Honey with a water content greater than 20 per cent is likely to ferment. This occurs if honey is harvested before the honey cells are capped over (or 'ripened'). Honey is hygroscopic and will absorb atmospheric moisture if it is not stored in an airtight container. It can also pick up taints and bad flavours very easily, so careful handling and storage are essential.

Harvesting Honey

When the moisture content has been reduced to around 18 per cent, honey cells are capped over with beeswax. Frames should not be extracted until at least 80 per cent of the cells are capped or the honey could ferment. Honey can be tested for capping by holding a comb horizontal and gently shaking it. If no honey flicks out, it can be extracted. If nectar flies out, the honey is unripe and likely to ferment if extracted.

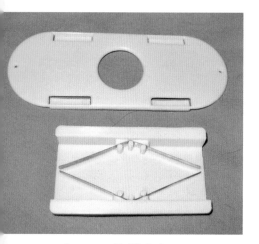

An unassembled Porter bee escape

Clearing Supers

Before taking the honey, the bees must be removed, or cleared, from the supers. The simplest way to do this is by using specially designed clearer boards. The most common, although I think now a bit old-fashioned, is the Porter bee escape.

The Porter Bee Escape

The Porter bee escape consists of two parts that slide together and then slot into a crownboard. The top has a central hole down which the bees pass. The bottom section has two pairs of flexible 'springs', which leave a gap just smaller than a bee space. A bee can push through it but the gap is too narrow for it to make the return journey. I like to check the spacing on these escapes by sliding off the top of the escape and gently adjusting the two pieces of wire with my thumbnail if the gap seems wrong. It should be 3–4 mm. Just imagine the size of a bee to get it right. This method will take a day or maybe even two to clear the bees out.

Did You Know?
A 'full' National super will hold around 9 kg (20 lb) of honey.

Rapid Escapes

More rapid methods, such as Canadian escapes, only take a few hours to clear bees from the super. One put on in the morning will be ready by late afternoon. These escapes come in several designs, again based on the principle that bees can pass one way but cannot return easily. One that works well is a lozenge-shaped plastic tray, roughly one bee space deep, fixed under the clearer board. The hole in the board allows bees through and they are 'funnelled' to the openings at the ends of the lozenge. However, if they are left too long, the bees will find the way back into the supers. Make sure you clean these types of escapes before use. If they get blocked they will fail, leaving the super full of bees when you return.

Top Tip
Make sure the escape is the right way up so the exit hole directs the bees into the hive. If it is upside down, you will get more bees than you know what to do with in the super you are trying to empty.

Sorting Supers

Remove the supers and sort out which are to be left and which are to be harvested. The bees leaving the full supers will need some space, so you may need to put an extra empty super under the clearer board. I generally sort out the supers I am going to leave for the bees, maybe the ones containing some unripe honey, and put those back on the hive. I then place the clearer board on top and add the supers I wish to clear of bees. Make sure that all the access points, such as the feed holes, are covered.

Bee-tight Hives

It is essential the hive is bee-tight when honey is being harvested. I often use masking tape to cover the joints between the boxes while the bees are clearing from the supers. It is also a good idea to make the entrances smaller, especially on weaker colonies, so the bees can defend their hives from robber bees and wasps. A slot a little under a bee space high (5–6 mm) x 100 mm long is suitable.

Top Tip

Do not leave the clearer board on the hive after harvesting, as the bees will fill the gaps with wax and propolis, making it useless.

Brushing Bees

Bees can also be brushed from the super combs for an instant result – especially if there are only a few colonies. Take an empty super box and stand it on a board. Lift the required supers from the hive and stand them across the upturned hive roof (to catch honey drips and bees). Brush the bees from each individual comb back into the hive and transfer the cleared combs to the empty super. Cover it with a cloth during the transfer because the bees will quickly locate their honey and want it back. Ideally, a helper moves each super away from the apiary as soon as it is cleared. Time the procedure for late afternoon so the bees have time to fly back to the hive but any excitement generated reduces once the evening comes.

Keeping Bees Out

It is also vital that the place you take the honey to for extracting is inaccessible to bees. Under no circumstances leave the kitchen door open! In less than ten minutes, every bee in the district will have found the honey and come to collect some. This is a truly alarming phenomenon, so beware.

Extracting Honey

Honey direct from the hive is quite warm. Warm honey is less viscous, comes out of the cells better and is easier to filter. To make extracting easier, do not leave cleared supers to cool down. Do it straight away.

Getting Ready

While you are waiting for the supers to clear, prepare the honey extraction area and equipment. A clean kitchen is quite acceptable for extracting honey on a small scale but, especially if honey is to be sold, a high standard of hygiene is essential. If you are thinking of a more commercial operation, the latest details are available on the website of the Food Standards Agency.

169

Uncapping a frame of honey

Arranging the Process

Try to arrange the workflow to move smoothly from one task to the next. Consider arranging the stages in a circle around where you will stand. Have the pile of full supers near the sink. Then organize the uncapping station, which should be convenient for extracting. After extraction, the empty supers can be stacked ready to receive the empty frames. Finally, there needs to be space for the filtering and bottling equipment.

Uncapping

The wax capping on the honeycombs must be removed before the honey can be extracted. This is most easily done with a sharp or serrated knife. One with a blade longer than the depth of the frame works best. The cappings are cut off into a clean container.

Collecting Cappings

As usual, there is plenty of equipment available to buy to collect the cappings (and honey that drips from them), but the simplest thing is a clean plastic washing-up bowl. To support the frame you are uncapping, use a 50 mm-wide length of wood long enough to span the bowl. Cut grooves into the underside at each end to fit the bowl's rim. To further stabilize the collecting process, cut a recess in the other side to hold the lug of the frame. Place the fashioned wooden frame support across the bowl and stand the frame lug in the recess. Then slice off the cappings. At this stage any irregular comb surfaces can be evened up. Don't worry if this means

170

Did You Know?

Cappings from honeycomb yield
the very best beeswax.

cutting off quite a bit. The bees will
repair the comb next year and you
can recover the wax and honey from
the bowl.

Protect the Floor

Work cleanly when extracting honey. The smallest
drops on the floor seem to turn rapidly into a sticky
carpet, one molecule thick, across the whole room.
Cover the workplace and floor with plastic sheet or
newspaper to catch the drips.

Honey Extractors

Honey extractors are expensive, especially on top of your other
start-up costs, so it is worth investigating whether your local
beekeeping association has extracting equipment available for
loan or hire. This service may see you through the first few years
until you decide what is best for your individual needs. Honey
extractors come in two types: tangential and radial.

Top Tip

Make sure the extractor you
choose will accommodate
the frames you are using.

The Tangential Extractor

The honey frames are placed in a wire cage, which rotates on a central spindle inside a barrel. The frames are placed at a tangent to the barrel, hence the name.

A tangential extractor

> ### Top Tip
> Turning an extractor too fast can break the combs, leaving lots of wax to be filtered from the honey. Much worse is the loss of precious drawn comb.

Tangential extractors are generally small, usually holding a maximum of six frames. This does mean they can be hand-powered, so they are normally cheaper than radial extractors. However, the liquid honey only spins out from one side of the comb so the frames have to be turned around, making the extracting process longer. The honey from the cells hits the side wall and runs down into a collecting well at the bottom. From here it is removed through a tap or 'honey gate', ideally into honey buckets (*see* image on page 179).

The Radial Extractor

With this design the frames are arranged like the spokes of a wheel. The honey spins out of both sides of the comb at once. Radial extractors will generally be larger and often motor-powered. With a motor-driven extractor, build up the speed gradually so that it can get 'balanced'.

A radial extractor

Balancing the Load

When loading the extractor, try to balance frames of similar weight either in a batch or opposite each other, otherwise the extractor will begin to move around alarmingly. If it seems a problem, fix strong castors to the base of the extractor legs.

Luc Viatour/www.lucnix.be

Filtering honey

This removes the strain on the central spindle. Honey in the bottom of the extractor needs to be removed periodically before it reaches a level that impedes the extractor's rotation. Be warned. Don't leave honey running out of a tap unattended. It flows silently and you won't be aware of an overflow until it's already all over the floor.

Filtering

Honey can be filtered and bottled straight from the extractor, but I like to filter it coarsely, using a conical tap strainer, run it into 13.5 kg (30 lb), food-grade honey buckets and tip the buckets through a filter into a ripener or settling tank. A double strainer with a coarse filter sitting on top of a finer one works well. The top filter catches the larger pieces of wax and the lower one removes finer particles.

173

A settling tank

The Settling Tank

This is simply a large food-grade vessel that will hold a decent amount of honey. Those commercially available are made of plastic or stainless steel and are designed to hold anything from 35 kg (77 lb) to 100 kg (220 lb). Once the honey is in the tank, it should be left for 24 hours or so to allow air bubbles to rise to the surface. With a little care, beautiful jars of honey are easily filled from the tap. The draw-off tap is placed below the scum and air bubbles on the surface of the honey so the clean honey can run out from underneath. Once the clearest honey stops, run the remainder into a bucket for home use. The whole harvest can be stored in buckets if it is not being used immediately.

Dealing with Wet Frames

When extracting is finished, return the 'wet' frames to the super. Take the supers back to the apiary in the evening and place them on top of the hives for the bees to clean up. You must place them above the crownboard with the feed hole open so the bees think the honey source is outside the hive and hurry to take it back inside. In a week or so, the bees will have cleaned out the remaining honey and the supers can be stored for the winter. Any bees still in the supers when they are to be removed can be cleared using bee escapes.

Top Tip

Place wet supers on the hive in the evening so the bees do not start robbing each other.

174

Dealing with Different Honeys

Honeys differ in colour and flavour depending on where the nectar was collected. Honey that granulates rapidly can be extracted by melting the honey and wax rather than using an extractor. Granulation can be controlled to produce soft-set honey. An alternative is to produce cut-comb honey. Heather honey is unique with a distinctive flavour and texture. Honeydew is another unique honey. Honey must be ripe before extraction or it will spoil and ferment. Honey for sale must comply with the appropriate regulations.

Melting Out Honey

Not all honey has to be spun out of the combs using an extractor. Some beekeepers leave oilseed rape (OSR or canola) honey to granulate in the comb and remove it later on in the year. Increasingly beekeepers, especially those working on a large scale, are not bothering with extractors but are extracting honey using a destructive comb removal method, where heat is used to separate the honey from the wax.

Granulated honey in the comb

Removing the Comb

This method requires comb to be cut from the frame (or scraped from the midrib). Comb is either scraped back to the midrib (which is rather time-consuming and tedious but preserves the foundation for next season) or just cut out from the frame and then chopped or mashed into small pieces. Some beekeepers cut the comb to leave a small starter strip for comb building next season.

Separating Wax and Honey

Once removed from the frame, the scraped or mashed comb is warmed gently so the wax is separated from the honey. Honey and wax liquefy at similar temperatures but when left to cool, the wax will solidify as a cake on top of the honey and can be removed.

A Simple Approach

There is very good commercial equipment available for this job but, for the small or cost-conscious beekeeper, it is possible to do this without expensive equipment. The mixture can be heated in any suitable container with a thermostatic control to give a steady end temperature of 64°C (147°F), which will melt both the honey and the wax. The disadvantages of melting out honey is that quality is likely to be reduced and comb has to be drawn out again each year. In addition, the melting process releases the pollen in the comb so that the honey can appear cloudy or murky. I would not use this method myself but then I do not live in an area where oilseed rape is grown.

Running Honey

If you have only a small amount of honeycomb, or bits and pieces of broken comb still containing honey that will not fit into the extractor, you can extract this honey using the run honey method. Again this is very simple but, be warned, it is very messy!

Run Honey Method

The comb is simply cut or scraped from the frame into a clean container where it is cut up very finely to break down the cell structure of the comb. The resulting mush is poured into a muslin cloth or a jelly bag so the honey drains into a receptacle below. Once the honey has drained completely, put the container in a warm place and wait 24 hours. The wax particles and foamy air bubbles will rise to the surface and you can skim them off, although they are perfectly edible. The honey underneath will be pretty clear but if you sell it you may want to filter it further.

Comb Honey

This is a popular way of presenting honey. If you choose this method, it is important to use thin and unwired foundation in the super frames. The lack of wire in the foundation allows the comb to be cut up

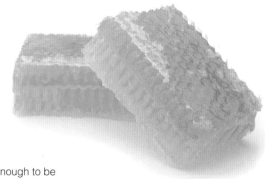

and the wax midrib of the comb will be thin enough to be eaten along with the honey. The cut-comb honey can be packed into attractive cartons available from beekeeping suppliers.

Heather Honey

Heather honey is the last commercial crop of the year. Bees must be out on the moors before the 'glorious' twelfth of August if they are to work this crop. The lateness of the heather crop and subsequent settling of the bees for the winter present their own problems and very often heather honey is collected simply as winter feed for the bees.

Thixotropic

However, heather honey lends itself especially well to cut-comb honey production because of its unique thixotropic (jelly-like) nature. This makes the honey difficult to remove from the comb. There are some neat gadgets that give each cell a little twiddle so that the honey liquefies and can be extracted normally, but they are unlikely to be economic unless a significant heather crop is collected regularly. Alternatively, comb can be scraped from the midrib and then pressed hard through a straining bag. Whichever way it is harvested, to my mind heather is one of best honeys ever. I like its slight bitterness, which offsets the sweetness.

Granulation

Granulation is the natural process of the glucose crystallizing out of the fructose solution. All honey will granulate over time and this is actually proof of local honey quality, because commercially pasteurized honey will not granulate. The rate of granulation depends on the glucose level and this varies according to the nectar source. As mentioned previously, honey from oilseed rape has a high glucose content and can even granulate in the comb. I was always taught

Honey buckets should be used for storing honey, so that it is easier to deal with granulation should it occur

that honey in jars should be considered perishable so, if it is not to be used straight away, it is best stored in honey buckets. If it granulates in the jar, it is nowhere near as easy to correct.

Fermentation

Fermentation of honey (unless you are making mead) is entirely undesirable. This occurs, making the honey smell and taste faintly alcoholic, if the water content of the honey is too high (more than 19 per

A refractometer

cent) – and honey absorbs water very readily. As long as the water content remains low, the natural yeasts present in the honey cannot multiply and cause fermentation. Honey is more likely to ferment when it has granulated. Microscopically, granulated honey is composed of crystals surrounded by a liquid honey solution – because the sugar concentrates in the solid crystallized elements, there is proportionately more water in the liquid element, leading to the risk of fermentation.

Preventing Fermentation

Avoid this by storing your honey in tightly closed containers. The moisture content of honey can be measured using a refractometer. In general, though, if the combs selected for harvesting were well sealed and the honey has been correctly stored, problems of high water content or fermentation are unlikely.

Soft-set Honey

One method of presenting granulated honey is to make soft-set or creamed honey, which gives it a lovely buttery consistency. The soft-setting method controls the granulation of the honey. If the honey has already set into a nice fine-grained consistency in your honey bucket, all you need to do is to gently warm it until it is soft, then stir it thoroughly, being careful not to introduce too much air into the mixture. This softened honey can then be bottled – it will set again but this time into the desired consistency, and will remain that way. It will not turn back into the very hard 'Geller' honey that bends spoons.

Did You Know?

'Creamed' describes the texture – not that it has been whipped to introduce air. You do not want extra air in your honey.

Controlled Granulation

If the honey does not have a pleasant granulation and feels 'gritty' or crunchy and coarse in the mouth then it needs a little extra treatment to make a fine-grained soft-set honey.

Liquefying Honey

To achieve this, the honey first needs to be completely liquefied. This is best done in a warming cabinet, which can be made or purchased but needs to be thermostatically controlled. To liquefy the granulated honey needs a gentle temperature of about 42°C (108°F) for 24–48 hours. Take care not to heat honey to a higher temperature or for longer than necessary.

Top Tip

When producing soft-set honey, you need to add 10 per cent of the fine seed honey to the bulk of the liquid honey in order to achieve the desired granulation.

Seeding Honey

Once the honey is completely liquid it needs to be 'seeded' with a suitably textured honey. Oilseed rape honey is ideal for this job because it is a very fine-grained honey that granulates quickly, but any fine-grained honey will do. If necessary, buy a jar. The seeding honey is softened slightly, then stirred into the liquid honey until it is thoroughly incorporated. The effect of this is that it will form a 'seed' or pattern and the honey will regranulate, taking on the new texture.

181

Labelling

If you sell your honey in the UK, you must comply with the Honey Regulations 2003, revised in July 2005 and amended in October 2007. These regulations, the Food Labelling Regulations 1996 and the Food (Lot Marking) Regulations 1996, give details of the required type and composition of your honey and how jars must be labelled. Rather than go into the details here, it is easiest to purchase labels from an equipment supplier as these will conform to the regulations.

Best Before Date and Lot Numbers

Honey for sale must carry a 'Best Before' date – most beekeepers go with either one or two years. You must also include a Lot Number, beginning with L, on your jars and keep records of your batches and where they are sold. Labels should also show the country of origin. This is frequently added to a tamper-proof label. Guidelines can be downloaded from the Food Standards Agency at www.food.gov.uk, and a leaflet is available from the British Beekeepers' Association at www.britishbee.org.uk/files/selling_honey_B10.pdf. Other countries will have broadly similar legislation but check out your local regulations.

Honeydew

Before we leave this section on honey, we should mention honeydew. This is different from honey because it is not processed from nectar but collected by the bees from the sweet exudates of plant-sucking insects. These insects excrete excess sugary liquids plus some additional insect-synthesized sugars. It is a darker, sometimes more bitter, product that is differently valued in different countries. I have had the most delicious honeydew in New Zealand, where the trunks of trees are turned black by the sticky activity of insects.

Beeswax

Beeswax is a wonderful honeybee by-product and one of nature's most amazing materials. Pure beeswax is a complex mixture of hydrocarbons, free acids and esters, and contains at least 300 different compounds. Produced in small scales from glands under the abdomen, bees use it to build their honeycomb.

The Value of Wax

People have long valued beeswax for its versatility. It is used as an ingredient in many commercial cosmetics, ointments and pharmaceutical preparations. It is also important in batik textile work and certain metal-casting and modelling processes, as well as wax foundation for beehives and for making candles. There are no more beautiful or delicious-smelling candles than those made from pure golden beeswax. There is something very special about burning a candle made with beeswax harvested from your own bees.

Different Ways of Obtaining Wax

Wax can be collected from combs, top-bar scrapings and cappings from honey extraction. It can also be recovered from the combs as they become too old for use in the colony. Cappings wax is probably the cleanest and purest you will get, and should always be kept separate from the rest.

Wax Extraction Methods

▶ **Melting in hot water:** Once the water cools, the wax hardens and can be removed.

▶ **Melting using a solar wax extractor:** The wax is placed under glass in the sun; it then melts and runs into a collecting container. Again, the wax hardens once the container is removed.

▶ **Melting by using a distillation method:** Here the wax is suspended in a filtering bag over a steam heat source. The melted liquid wax runs through the bag into a container placed underneath.

Extracting Wax from Cappings

Let us consider making beeswax using the cappings from honey extraction. First, wash the cappings with plenty of clean, soft water. Filtered rainwater is fine.

Melting in Hot Water

If you only have a small quantity of cappings, they can be melted in a microwave. Put some clean, soft water into a heatproof glass bowl and add the washed cappings. Heat the cappings in short bursts. Avoid boiling the water as this will mix dirt and debris into the wax and darken it. Wax melts at 64°C (147°F). Alternatively this can be done on the hob. Again, do not boil the water; a gentle below-simmer heat is enough. When the wax is melted let it cool slowly. The wax will solidify on top of the water and can be removed. Dirt on the underside can be scraped off but the wax will need filtering before it is used or sold.

A solar wax extractor

Using a Solar Wax Extractor

Personally, I use a solar wax extractor. I find it has the virtue of simplicity, for which I am a great enthusiast, as well as being free. Again, although they can be purchased, they are very easy to make. Mine is made of a caravan window hinged onto an insulated wooden box. The box is lined inside with a metal tray, angled and shaped at the bottom to direct the melted wax into a container. On a decent sunny day, the wax melts quickly. I push the cappings into the legs of old tights, which go straight into the extractor. The melting wax runs into the container while the dross is left behind in the filters (tights) and can be burned. It makes very nice-smelling firelighters. When the wax is removed from the extractor, it will solidify.

Using Steam Extractors

There are also really good methods of extracting wax using steam. For those with plenty of money, a commercial wax extractor is lovely. Otherwise, wax can be hung in a filtering bag in a metal container. The bag is hung directly over a bowl that is floating on a quantity of boiling water. The whole contraption needs to be placed over a heat source to produce steam. The steam rises up around the wax cappings in the bag and melts them. The liquid wax runs though the filter bag and into a bowl floating directly underneath. Although it sounds a bit fussy, it works very well.

A steam extractor

185

Filtering Beeswax

Once the wax has been turned into a solid cake, it will need further cleaning by filtering. To do this it must first be liquefied again. In the past, I used the textbook double-boiler method, making a *bain-marie* arrangement of a bowl inside a saucepan to melt the wax, but these days I melt the wax in a microwave until it is liquid. Once liquefied you need to move quite fast as it will quickly cool and solidify. Pour the liquid wax through a filtering cloth into a jug. Materials that work well for the purpose include old sheets, T-shirts and J-cloths because they are cheap and disposable. When you have finished, use the wax-clogged leftovers for firelighters.

Moulds

Being well prepared and having two pairs of hands are both infinitely helpful when working with wax. It is perfectly possible to use the wax straight away. Otherwise pour it into moulds and let it set. Plastic containers used for soft butter or margarine are ideal. The wax will shrink away from the surface as it solidifies, and the handy blocks can be removed. One-ounce moulds are available from beekeeping suppliers and will help if you need measured quantities of wax for a recipe.

Top Tip

Wax processing is messy, so dedicate the utensils used to this purpose alone.

Exchange for Foundation

If you do not want to be creative with your filtered beeswax, it can be exchanged for new sheets of foundation at several of the larger beekeeping equipment suppliers. Some beekeeping associations collect wax from members and organize a bulk deal with an equipment supplier, with members receiving an agreed number of sheets of new foundation in exchange.

Propolis

Propolis is a sticky, resinous substance collected from different kinds of trees and plants. The bees mix in substances derived from pollen and active enzymes to form a potent, naturally therapeutic substance. Over 180 distinct compounds have been identified in propolis with researchers expecting to find more. The concentrated plant-derived essential oils are thought to be mainly responsible for the product's therapeutic properties.

How Bees Use Propolis

Honeybees use propolis to keep their homes dry, cosy and hygienic. It is used to seal any cracks or gaps where microorganisms could flourish and to decrease the size of nest entrances, making them easier to defend. To collect it, bees bite off scraps of plant resin and pack them into the pollen baskets on their hind legs. They only collect it when the temperature is above 18°C (64°F), and each bee can only carry about 20 mg in one journey, so propolis gathering is a slow business.

Harvesting Propolis

Propolis can be harvested by placing a perforated plastic screen, similar to a queen excluder, on top of the hive. The holes in the plastic are small,

Fresh propolis

187

so the bees fill the gaps with propolis. The screen is placed in the freezer, with the frozen propolis being collected by flexing the screen. About 50 g per hive per season can be harvested this way.

Using Propolis

Propolis has anti-fungal, antiviral and antiseptic properties. It is even being investigated for anti-cancer properties. Health-product companies use it to make throat lozenges and propolis tablets or syrups. It has many uses, both internal and external, and can be used either fresh or as a tincture. Fresh or powdered propolis can be chewed and swallowed for all types of stomach problems and sore throats. A small piece of propolis applied directly to the source of an aching tooth will relieve pain. It can be used for boils, ringworm, fungal infections and all sorts of wounds and sores. Propolis is always most effective straight from the hive.

Propolis Tincture

The most common non-industrial extraction process is to soak the propolis in 70 per cent ABV alcohol or higher for three weeks to make a tincture. This gives the maximum extraction of the most significant active elements. Since pure alcohol is strictly controlled in the UK, vodka is the most commonly used alternative.

Making a Propolis Tincture

For a propolis tincture, one litre of alcohol is required for each 250 g of propolis. The propolis is soaked in water for 3–7 days to clean and soften it. It is then dried gently and soaked in alcohol for 1–3 weeks, shaking the mixture every day. The resulting solution is filtered through a fine filter such as a coffee filter. The medicinal part is the filtered liquid. This should be kept in a dark glass bottle in a cool place. The tincture will keep for a long time because alcohol is an excellent preservative.

Checklist

▶ **Honeys of all colours:** Honeys have different colours and flavours depending on the plants from which they were derived.

▶ **Mechanical extraction:** Honey extractors come in two types – tangential and radial.

▶ **Heat extraction:** Honey can be extracted by melting the comb carefully under controlled conditions; it is critical that honey is not overheated.

▶ **Creamed and soft-set:** Honey's natural granulation process can be controlled to produce a soft-set honey.

▶ **Check for ripeness:** Unripe honey will ferment after extraction.

▶ **Check requirements:** Honey for sale must conform to the relevant legislation.

▶ **Extracting beeswax:** Beeswax can be melted over hot water, by steam or in a solar wax extractor.

▶ **Other useful honey by-products:** Propolis is a sticky resin produced by trees and used by honeybees to block gaps in the hive.

▶ **Medicinal:** Honey and propolis have anti-fungal, antiviral and antiseptic properties.

Pests

And

Problems

Stockmanship

Caring for bees requires stockmanship, often referred to as 'reading the bees'. This is learned over time and is based on keen observation plus understanding. Looking is no good without seeing, and seeing is based on knowing, so theoretical knowledge is important. Forewarned is forearmed, so here are some problems to look out for.

Natural Controls

The warm, humid environment in a beehive is perfect for spreading disease. It is a tribute to their natural disease control mechanisms – their immune systems, biology and social behaviour – that honeybees are so resistant to disease. Different pathogens challenge colonies in different ways and many diseases

and parasites have co-evolved with the bees over the millennia. Individual colonies vary in their resistance and it is this genetic element of disease resistance that selects the colonies that will successfully adapt to their location. Locally adapted bees are likely to be increasingly important as changing weather patterns bring additional challenges.

Good Hygiene
Strict sanitary conditions are vital. Bees are constantly cleaning both themselves and their nestmates. Debris is removed from the nest with great alacrity and, where this is not possible, it is

covered in propolis to isolate it. Honey and propolis are natural antibacterial agents and the supersaturated sugars in honey inhibit bacterial or yeast growth. Cells are cleaned carefully between brood cycles and only used if they meet the bees' strict hygiene standards.

Young healthy worker bee eggs and larvae

Removing Disease

Bees quickly recognize sick larvae and try to remove them from the hive before the disease can spread. This results in the 'pepper-pot' or patchy brood pattern, which is one indication that all is not well. Feeding or heavy nectar flows can reduce the incidence of disease by diluting any pathogens present. Heavy nectar flows also keep potentially infective foragers away from the bees in the hive and wear them out so they die away from home. Conversely, restricted foraging, because of poor weather or too many bees in one area, exacerbates the virulence of any pathogen. Vigorous queens and plentiful forage are essential prerequisites for healthy, productive colonies.

Top Tip

Eggs hatch after three days; cells containing worker larvae are sealed six days later, so there should be twice as many larvae as eggs if the queen is laying consistently. It takes 12 days for a worker to emerge as an adult, so there will be twice as many sealed cells as those containing larvae.

Responsibility of Care

Beekeepers often dread any mention of disease but, by deciding to keep bees, you have accepted the responsibility to care for them in the best possible way. Happily, diseases are rare but it is important to understand the difficulties that may occur, and develop strategies to reduce the chances of problems arising.

193

Assessing the Health of a Colony

The first essential is to recognize what is healthy and what a colony should look like at a given point in its development. Recognizing eggs and a good brood pattern will tell you the queen is present and the larvae are developing as expected.

Signs of a Healthy Brood

Brood of different ages should be present in concentric circles on a brood comb. The queen will be laying in virtually every cell so they will contain an egg, or a developing larva, or they will be sealed. Eggs should sit in the centre of the base of the cell. Larvae should be a glistening, pearly white and lie comfortably curled up in the cell. Sealed brood is the dominant feature of an established brood nest. Healthy sealed worker brood has uniform, slightly domed cappings. Domed cappings on drone brood are more pronounced.

Healthy hatching brood

Dwindling and Weak Colonies

Colonies that are not expanding or are weaker than expected indicate an underlying problem. With more than one colony you can compare development to assess their health. Causes of dwindling and weak colonies include:

▶ Lack of nectar, honey stores or pollen
▶ Problems with the queen
▶ Pests, parasites or disease

194

Beekeeping Problems

Several pests, parasites and diseases can cause colony losses, the worst being *Varroa destructor* (*see* pages 201–05). Laying workers or a drone-laying queen will seriously depress the colony. Specific honeybee diseases can be divided into brood diseases and adult diseases. American Foul Brood (AFB), European Foul Brood (EFB) and minor brood ailments, including chalk brood and sac brood, affect the larval stages. Acarine and Nosema infect adults. A wide range of pests and predators can also cause problems. Some pests and diseases are notifiable by law.

Notifiable Diseases

Two diseases and two pests are currently notifiable in the UK. AFB and EFB are serious brood diseases. The two pests, Small Hive Beetle (SHB) and the tropical mite Tropilaelaps, are currently notifiable but not yet present in the UK. If you suspect any of these are present in your colony, you must, by law, inform the relevant government authorities. AFB is considered a very serious disease worldwide and is subject to statutory control in most countries.

Small adult hive beetles are just visible amongst the bees here

195

Brood Diseases: American Foul Brood (AFB)

The two most serious brood diseases are AFB and EFB. The names only signify where they were first identified, and both are found worldwide. They are both caused by bacteria. The minor brood disease chalk brood, caused by a fungus, is very common.

Progression of AFB

AFB is caused by the spore-forming bacterium *Paenibacillus larvae*. Typically, the young larva ingests AFB spores along with its food. These germinate in the gut, invade the body tissue and multiply rapidly.

Eventually the infection overwhelms and kills the larva and its remains collapse into a sticky 'goo' that gradually dries up into a scale on the lower side of the cell. This sticks so tightly that the bees cannot remove it, however hard they try. These dead larval remains contain billions of bacterial spores that are passed on to new larvae. Eventually, larvae die more quickly than they can be replaced and the colony dies out.

Inevitable Death

AFB invariably kills an infected colony, normally over one season or so, and it leaves behind a legacy of infected combs to spread the disease. The spores are resistant to extremes of heat and cold and are unaffected by disinfectants. They can persist in the hive and combs, and on other equipment, for up to 50 years.

An American Foul Brood (AFB)-infected comb – the white mummies are chalk brood

Keep Vigilant

As infected larvae die, the cell cappings become dark, sunken, wet-looking and perforated; very different from the nice slabs of even-coloured cappings over healthy larvae. You can easily miss the dried scales of larval remains in empty cells, especially if the comb is darkened from generations of brood rearing. Do not think just because colonies are strong that they cannot have AFB. A weak colony or one that has died from foul brood may be robbed by its larger, fitter, more aggressive neighbours, taking spores home to begin the cycle again.

The 'ropiness' test

The Ropiness Test

A simple field test for AFB is the 'ropiness' test. Select what looks like an infected cell and push a matchstick into it. If it is infected, when you withdraw the matchstick, it will pull out the larval remains in the form of a mucus-like 'rope' for 10 mm (⅜ inch) or more.

Top Tip

To inspect for AFB scales, stand with the light coming over your shoulder. Hold the frame at an angle of roughly 45° and look down into the cells. You will see the scales on the lower surface.

Disease Inspection

Especially in the early stages, there will only be a few scattered infected cells, making the disease harder to detect. This is why you should shake the bees from the comb for a disease inspection so you can check the cells more easily. The queen will avoid cells that are not cleaned properly, which often gives the brood a patchy appearance, although this can also typify other diseases.

Signs of AFB

▶ Dark, sunken, perforated cappings which may look greasy
▶ Dried larval remains forming tightly adhering scales in empty cells
▶ A patchy appearance where numerous random cells have been uncapped

Control of AFB

The best way to deal with AFB is destruction of the colony. If the disease is confirmed in the UK, the bees are compulsorily destroyed under the supervision of your bee inspector, along with any appliances and equipment that cannot be sterilized. All the frames, combs, bees and honey are burnt and the remains buried. Where practical, hives and appliances are sterilized by blowtorch to kill the spores.

Brood Diseases: European Foul Brood (EFB)

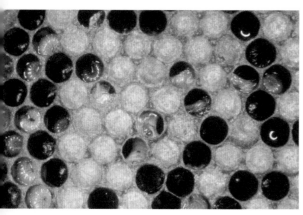

A European Foul Brood-infected comb

EFB is widespread in some parts of the UK but rare or unknown in other places. It is caused by the bacterium *Melissococcus plutonius*. EFB infects the larval gut and the larva starves to death shortly before the cell is sealed, usually at 4–5 days old and mostly in early summer when colonies are expanding rapidly. Characteristically, EFB-infected larvae twist around uncomfortably or stretch out in their cells. The sick larvae become discoloured, often yellowish brown, and look 'melted'. When the larva dies, it dries into a brown scale, similar to that of AFB, but easily removed by the bees.

Symptoms of EFB

▶ Infected larvae lie uncomfortably in the cells

▶ Larvae appear 'melted' and discoloured

▶ A severe infection shows a patchy brood pattern

▶ Secondary bacterial infections may cause an unpleasant smell

▶ Larval remains do not 'rope' as with AFB

Identifying EFB

EFB bacteria act as parasites killing the larvae by competing for their food. As the bees clean out the casualties, the clinical signs of disease disappear and bacterial levels in the combs are greatly reduced. This means that disease signs obvious at one inspection may have completely vanished by the next. However, once a colony has had EFB, a reservoir of bacteria remains in the comb with the potential for re-infection. EFB can occur at any time of year but it is most obvious in the spring, when the food supply only just meets the needs of the larvae.

Did You Know?

There are lateral flow devices for testing for AFB and EFB. A larva with suspected symptoms is placed in an extraction bottle and shaken. A sample is withdrawn and a drop placed in the well of the device. The result can be read after 1–3 minutes. A blue control line appears and a second blue line indicates disease.

Dealing with EFB

There are three possible courses of action if EFB is found in a colony: colony destruction, treatment with the antibiotic oxytetracycline (OTC) or shook-swarm removal of the brood. However, antibiotics do not cure the disease, only suppress it. Antibiotic treatments should always be combined with comb-change techniques to remove remaining bacteria. In England and Wales, this is supervised by the National Bee Unit Inspectors. In other places, seek advice from the relevant Inspectorate.

Minor Brood Diseases

The symptoms of chalk brood and sac brood, together with some queen-related problems, can easily be confused with those of the more serious brood diseases. Sac brood is described in the virus section (*see* page 210).

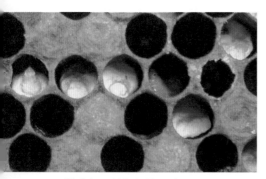

Chalk-brood-infected brood

Chalk brood 'mummies' on a hive floor

Chalk Brood (*Ascosphaera apis*)

This common fungal infection kills the larva after the cell has been sealed, again causing a patchy brood appearance. The larva turns chalky-white as the fungus develops. These 'mummies' may be obvious on the floor of the hive. Some bees are more susceptible to chalk brood than others. A severe infection may be reduced or eliminated by re-queening the colony. Chalk brood is rarely serious, but the fungus does produce millions of spores that stick to combs and adult bees.

Because the spores are viable for several years, replacing old brood combs on a regular basis will reduce the incidence of the disease.

Did You Know?

Disease at low levels often goes unnoticed, with the result that it is easily spread amongst colonies by the beekeeper moving combs about.

200

Varroa

The parasitic mite *Varroa destructor* is a very efficient killer of honeybee colonies. The mite also acts as a vector for viruses and other diseases. It crossed the species barrier from the Asian honeybee (*Apis cerana*) to our Western honeybee (*Apis mellifera*) and has spread rapidly throughout the world using human channels of international trade – we have become very efficient at globalizing bee diseases.

Varroa destructor mite on honeybee host

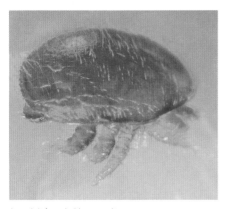

An adult female Varroa mite

Monitoring and Control

Proper monitoring and treatment are the keys to successful control. Spread between hives maintains mite populations in every colony in the area. The mite has already evolved behavioural means of surviving when its host colony dies. It lowers the defensive reaction of the bees to robbers. The dying colony allows robber bees in, which then carry the Varroa mites back to their own hives. It has been estimated that 60 per cent of mites can survive the death of the host colony in this way.

201

The Varroa Lifecycle

Varroa mites feed on the 'blood' (or haemolymph) of both larvae and adult honeybees. The whole lifecycle takes place within the honeybee colony and consists of a phoretic stage, where the mites live on the adult bees, and a reproductive phase that occurs inside sealed brood cells.

The Reproductive Phase

To breed, the female mite enters a brood cell at the point of sealing. Once the cell is capped, she starts to lay eggs. The first egg is always a male while all the subsequent eggs are female. The single male mates with his sisters then dies. The number of female mites emerging from the cell depends on the honeybee larva's development time. Only mites that are mature by the time the cell is opened by the emerging bee will survive. They therefore prefer drone brood, which has a longer development period, allowing more mites to reach maturity.

A worker bee with deformed wing virus (DWV), which can be spread by Varroa mites

The Phoretic Phase

During the winter, or other broodless periods, the mites can live for up to six months feeding on adult bees, using their sharp mouthparts to pierce the bee's body between the segments. Mites can only survive a few days without feeding on bees, so they cannot survive on empty combs or equipment.

Identification of Varroa Mites

Varroa mites are oval and flattened and are relatively easy to spot with the naked eye. A heavily infested colony is likely to show brood abnormalities as well as adult bees with stunted wings and abdomens. These signs are often termed 'varroosis'. A few mites will not cause much damage to a colony. The potential for harmful effects on the bees increases dramatically as mite numbers rise; the currently accepted threshold for harm is about 1,000 mites.

Monitoring Mite Numbers

Regular checking of mite levels enables beekeepers to determine the optimal time to apply treatments, to check the effectiveness of the treatment used and to recognize new or unexpected mite reinfestation. Ideally, you should monitor colonies four times during the season: in early spring, after the spring honey flow, around honey harvesting time and in late autumn. The two simplest methods are drone brood uncapping and counting natural mite mortality.

Drone Brood Uncapping

Varroa mites are easily seen against the white bodies of drone pupae. This can only be done in strong colonies in the summer and is unlikely to detect low infestations. Use the following method:

▶ Select an area where the drone pupae are at an advanced (pink-eyed) stage

Use of an uncapping fork for Varroa mite detection

- Slide the prongs of an uncapping fork under the brood cappings
- Lift out the drone pupae to check for the pinhead-sized brown mites
- Examine about 100 drone pupae
- Estimate the number of pupae with mites
- As a rough guide, if more than 5–10 per cent of the pupae are infested, mite numbers are high and colony collapse is likely by the end of the season

Natural Mite Mortality

Counting natural mite mortality is a sensitive method capable of detecting low infestations and can be done at any time of year. If the hive is not already on a special Varroa floor, replace the solid floor with an open-mesh Varroa floor. These have a removable sampling tray underneath to collect floor debris. For new beekeepers, mesh floors should be considered standard equipment. Remove the collecting tray during the summer, except for the monitoring periods, to prevent an excessive build-up of debris and consequent infestation by wax moth.

Checking the Varroa floor

Natural Mite Mortality Method

- Insert the debris-collecting tray and leave in place for 3–5 days
- Remove, and count the number of mites found in the floor debris
- Divide the total number of mites by the number of days the tray was in place, to get an average daily mite drop
- The average daily mite drop depends on the Varroa population in the colony and the amount of emerging brood

▶ A colony with an average daily mite drop of 0.5 in winter/spring is, without treatment, likely to collapse by the end of the season

▶ Also treat if you have an increasing average daily mite drop, e.g. 6 in May, 10 in June, 16 in July, 33 in August, 20 in September

Integrated Pest Management

None of the medications or biotechnical methods currently available is totally effective, and it is impossible to eradicate the mites completely. At best they can be kept at levels that will not damage the colony. Integrated Pest Management (IPM) uses a combination of methods to gain the best mite control. Currently, there are two control methodologies: chemical varroacides and biotechnical methods. The whole topic is covered comprehensively in a book, *Control of Varroa*, from the New Zealand Ministry of Agriculture, or in the 2009 FERA booklet, *Managing Varroa*, which is available as a free download from Bee Base.

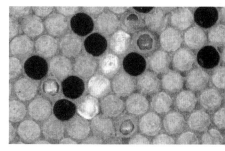

'Bald brood' caused by and showing greater wax moth larvae activity: allowing debris to build up on the mite collecting tray can encourage wax moth infestation

Top Tip
Plan control methods to keep the numbers of Varroa mites as low as possible in autumn to prevent damage to the larvae destined to become the 'winter bees'.

Timing Treatments

Varroacide treatments are likely to feature in the early autumn as part of winter preparations, while biotechnical methods are used during the active season. Varroa treatment is most effective when timed to protect the winter bees. Mite numbers are at their greatest in autumn when the smallest number of larvae are being raised, so lots of mites want to feed and breed on far fewer larvae. These larvae need to develop into adult bees in prime condition if they are to survive the winter, so it is essential the mites do not harm their development.

Mite Control Methods

Authorized varroacides are effective but relatively expensive. They may potentially leave residues in honey or wax, and mites can develop resistance to them. Other chemical treatments have variable effectiveness but are usually relatively cheap and may offer options not provided by authorized substances. They can also potentially leave residues. Biotechnical methods generally involve trapping and destroying mites in the comb.

Using a thymol treatment

Authorized Varroacides

Originally, beekeepers controlled the mites using synthetic pyrethroid varroacides (Bayvarol® and Apistan®). However, the mites quickly became resistant to the chemical, introducing the need for IPM techniques. Generic authorized products in the UK are Apiguard®, ApiLife Var® and Thymovar®, all based on thymol and other essential oils. Hive cleansers are Varroa-Gard®, Exomite® and Vitafeed Gold®.

Thymol Treatment

For thymol to be fully effective, the daytime temperature must be around 15°C (59°F). Although it can be applied at any time of year, it is most often used in autumn but treatment must be completed before the ambient temperature drops too low. Thymol is smelly, so honey must be harvested before it is applied. Do not feed bees when using thymol as it encourages robbing. Give any winter feed immediately after the thymol treatment is completed. Full instructions are available on the pack.

Generic Acaricides

Top Tip

Thymol treatment is usually combined with an oxalic acid application during the winter to ensure sufficient efficacy.

Organic acids (formic acid, oxalic acid and lactic acid) are also used as Varroa treatments (or acaricides) to kill mites. Their use is a legal grey area but, as long as no claims are made for treating Varroa, and government-monitoring schemes do not detect a residue build-up in honey or wax, their use is not prevented. You may find government literature offers advice on these compounds.

Oxalic Acid

Oxalic acid is most generally used in combination with thymol-based treatments. It only kills mites on the adult bees. It damages open brood and cannot kill mites in sealed cells so it is only used during a broodless period, most often winter. The solution is quite poisonous to humans so should be stored safely and suitable precautions taken when using it. It is best purchased ready-made to get the correct concentration for use.

Treating with Oxalic Acid

▶ Fill a 50 ml syringe with room-temperature oxalic acid solution

▶ Open the hive to expose the clustered bees

▶ Trickle 5 ml of solution along each seam of bees

▶ Close the hive

Biotechnical Methods

Biotechnical methods encourage mites into specific brood cells, which are removed when they are sealed with the mites trapped inside. They utilize the fact that mites are attracted to breed in drone brood.

Oxalic acid trickling treatment

207

The simple removal of three or four sealed drone combs in spring potentially reduces the autumn mite population by 50–70 per cent. These methods use no chemicals and can be combined with normal husbandry procedures. This technique can be used throughout the spring, once plenty of drone brood is present, to remove mites from colonies with low infestations.

Sacrificial Drone Brood

▶ Insert a drawn super comb in the brood box next to an outside frame containing brood
▶ The bees will build natural drone comb along the bottom of the frame to fill the gap
▶ When it is filled with drone larvae, the mites will occupy the cells
▶ When the brood is sealed (in 9 days), this comb can be cut off, complete with the trapped Varroa mites
▶ The wax can be retrieved as normal
▶ The super comb can be replaced for another two or three cycles

Varroa Control Strategy

▶ Assess mite levels regularly
▶ Frequency of treatments depends on the growth of the mite population and the risk of re-infestation from neighbouring colonies
▶ Mix acaricides and biotechnical methods in an IPM strategy
▶ Treat with varroacides when necessary
▶ Avoid using varroacides during the nectar flow
▶ Treat to maximize the survival of winter bees
▶ Combine different treatments, with different modes of action, to reduce the chance of resistance developing

Top Tip
Always follow label instructions. Do not be tempted to think that by applying twice as much, and/or leaving it in the hive for twice as long, it will be twice as effective. This will be building mite resistance to the product.

Viruses and CCD

Increasingly, new problems are arising from the synergistic effects of varroosis with other pathogens, especially viruses. Viruses can survive only as part of the host's cells, which makes them very difficult to eradicate without killing the host – that is, the honeybee. Until the arrival of the Varroa mite most beekeepers took little notice of bee viruses because they rarely caused a problem. Colony Collapse Disorder is not a virus, but viruses may be one of the many factors involved in CCD.

Associated Viruses

Certain viruses are associated with specific bee diseases such as varroosis. The paralysis viruses may show up as crawling bees on the ground, while deformed wing virus produces stunted wings.

Preventing Viruses
- ▶ There are no treatments for viruses, so treat the related diseases, mainly Varroa and Nosema (*see* page 211)
- ▶ Keep Varroa populations low by careful monitoring and treating with an effective varroacide as required
- ▶ Test for Nosema in the autumn and feed carefully to avoid dysentery, which spreads Nosema spores
- ▶ Do not import queens, as the best queens are always those that survive reliably in your own local area
- ▶ After the death of a colony, hive scorching and sterilization of combs with acetic acid removes any traces of disease

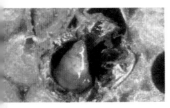

A larva with sac brood protrudes from its cell

Sac Brood

A larva with sac brood virus fails to pupate because it cannot shed its final larval skin. The dead larva stretches out in the cell in a fluid-filled sac forming a distinctive gondola or 'Chinese slipper' shape. It turns yellow then black, finally drying into an easily removable brown scale. Sac brood rarely affects many larvae and generally clears up without treatment.

Colony Collapse Disorder

Colony Collapse Disorder (CCD) is essentially an American phenomenon, although losses have been recorded elsewhere. Notable levels of colony deaths usually coincide with years when a combination of negative factors has affected honeybee colonies simultaneously. These can include poor weather, pollen shortage, ignorant use of pesticides and novel or endemic diseases.

Symptoms of CCD

▶ The majority of adult bees leave the colony over a short period

▶ These bees are not found near the hive, as would happen if they had been poisoned

▶ The queen remains in the hive with a small number of workers

▶ A dead CCD colony is not immediately robbed out by other bees, as would normally happen

▶ There is an unusual delay before wax moths invade the dead colony

Research

Much research is being conducted into the causes of CCD, and it is clear is that there is no single cause. Further research needs to be undertaken on the possibly harmful effects on honeybees of the use of neonicotinoid pesticides and gentically modified crops – as yet it is not at all clear.

210

Adult Bee Diseases

Adult bees can also be afflicted with diseases. The most common are Nosema and Amoeba, which affect the gut, and Acarine mites, which affect the breathing tubes, or trachea. Varroa mites also affect adult bees but their effects are worst on the developing larvae.

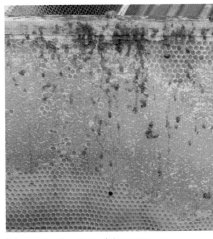

A frame showing signs of dysentery, which can be caused by Nosema

Nosema

Nosema is the most significant, universal bee disease. Until recently, the only causative agent known was *Nosema apis,* a specialized parasitic microsporidian fungus, but recently bees have been found to be infected by *Nosema ceranae*, which originated in the Asian honeybee, *Apis cerana*. This is proving to be a more aggressive form of the disease and another example of a globalized bee disease.

The Effects of Nosema

The spores survive on combs, frames and boxes for many months. When house bees clean up the hive they accidentally ingest the Nosema spores, which multiply in the gut. This can cause dysentery, although not all bees with dysentery have Nosema. Dysentery is manifest by brown faecal streaks on the inner cover, comb and other surfaces. Nosema shortens the bee's life. Consequently, the colony does not build up well, and dwindles or dies out in the spring.

211

Identification of Nosema

Nosema can only be positively identified using a compound microscope, although you may have strong suspicions if your colonies don't build up in spring. Your local association may have a microscopist who can do this examination for you.

Treatment for Nosema

The antibiotic fumagillin, available as Fumidil B®, kills the active stage of Nosema in the bee. If necessary, it is added to the winter feed. Mix the powder with some sugar and make this into a paste before adding it to the bulk of the syrup. Or try Vitafeed Green® in spring. Because Nosema spores can remain viable for 12 months and the disease is readily spread through the use of contaminated combs, it is essential to replace brood combs with clean comb or foundation periodically.

Amoeba

Amoeba, or *Malpighamoeba mellificae*, is a protozoan pathogen of adult bees that infects the bee's excretory organs. Amoeba often occurs with Nosema. It can only be identified by microscopic examination as it has no outward symptoms and does not currently cause any problems for the beekeeper.

Acarine or Tracheal Mite

The mite, *Acarapis woodi*, is also called the Tracheal Mite because it invades the bee's breathing tubes, or trachea, and breeds there. This shortens the bee's life and slows colony development. A severe infestation can kill the colony. Acarine is identified using a low-powered dissecting microscope. There is strong evidence of genetic resistance to Acarine, so, with no legally available treatment (in the UK), it is probably best to let susceptible colonies die out, or re-queen them with a queen that shows more resistance.

Serious Pests

Our honeybees also have to contend with a number of serious pests. Fortunately, the two notifiable pests, *Tropilaelaps* and Small Hive Beetle, have not yet been found in the UK. However, rodents, wasps and other insect robbers, and wax moths are major pests that easily destroy combs and colonies. Add birds, other wildlife such as badgers, livestock and human vandals, and you can see that much care needs to be taken to protect the bees.

Tropilaelaps

Tropilaelaps is a parasitic mite species currently restricted to tropical or sub-tropical regions – but, if it arrived, it could survive in the UK. It has been declared a notifiable pest. If you suspect its presence, you have a legal duty to inform the authorities. It is primarily a parasite of the giant Asian honeybee, *Apis dorsata*, although, like Varroa, it has transferred to *Apis mellifera*. It lives in brood cells. It is smaller and more rectangular than Varroa. It is detected in a similar way by uncapping drone brood and inspecting the pupae.

A *Tropilaelaps* mite

Small Hive Beetle

Small Hive Beetle (*Aethena tumida*) originated in Africa but has already become a serious pest in Australia and the USA. It is notifiable in the UK.

An adult Small Hive Beetle

213

The 5–7 mm long black beetles can be identified by their clubbed antennae, while the 10–11 mm long beige larvae are recognized by spines on their backs and three pairs of legs behind the head. The female beetle enters the colony and lays eggs in nooks and crannies. The larvae eat bee eggs and brood, spoiling the combs by leaving slime all over them and causing the honey to ferment.

Small Hive
Beetle larvae

Identification of Small Hive Beetle

▶ When you open a colony, look for beetles running away from the light
▶ Remove the supers (and upper brood chamber of a double-brood colony), stand them in an upturned roof and cover with an inner cover
▶ After a few minutes, lift the boxes and look for beetles in the roof
▶ Examine all the brood combs carefully and check the floor for larvae
▶ Look for clusters of eggs in cracks and crevices
▶ Use a corrugated cardboard beetle trap to check whether they are present

Lesser wax moth comb damage

Wax Moths

Unchecked wax moth infestations cause serious damage to both stored combs and live colonies. The lesser wax moth (*Achroia grisella*) causes greater problems in stored comb. The greater wax moth (*Galleria mellonella*) is a highly destructive pest that can very rapidly reduce both stored combs and weak colonies to a disgusting mess of webbing and excreta. As always, prevention is better than cure. The most important thing in the apiary and the equipment store is to keep things clean. Open-mesh floors can be a wonderful haven for wax moths, so keep them clean.

Ways to Control Wax Moth

▶ Store piles of empty supers on a stand outside; put mouse-proof mesh (such as a queen excluder) at the top and bottom of the pile and cover with a watertight roof

▶ When a colony is ready for winter, store supers on the hive, above the crownboard with the feed hole open

▶ Wrap combs in plastic and store them in a deep freeze

A deceased greater wax moth

Biological Controls

Certan® biological control for wax moth is safe for both stored comb and live colonies. It is a suspension of spores of the bacterium *Bacillus thuringiensis* (or BT), which is sprayed onto each comb. The herb wormwood (*Artemesia absinthium*) is reputed to deter moths. Hang small bunches of the fresh herb throughout the super stack. For the future, research with a parasitic wasp called *Trichogramma* looks promising.

Silver Lining

Wormwood can deter wax moth

Every cloud has a silver lining, and the wax moth is sometimes called the Beekeeper's Friend because it reduces old comb to threads very rapidly so they become unattractive to bees. If the comb contained disease, the disease organisms will be destroyed along with the comb. A new swarm of bees will clear everything out and build fresh, clean wax.

Top Tip

Avoid wax moth damage by keeping your colonies strong so the bees can keep an infestation at bay.

215

Rodents

Mice and rats can get into colonies and into stored equipment. They will generally only be able to enter occupied hives when the bees are not active. Mice can cause havoc by gnawing wooden parts, eating stores, chewing holes in comb and bringing in nesting materials. Shrews are insectivorous and can kill a colony by eating the worker bees while in their winter cluster. Rats pose an additional problem because they carry Weil's disease. If they get into your equipment, replace combs with new foundation and scrub other hive parts thoroughly with a strong disinfectant before re-use.

Mice can cause a lot of damage

A mouseguard in place

Keeping Out Rodents

Shallow entrances or mouseguards will keep out mice and rats. A mouseguard is a metal strip with 10 mm (3⁄8 inch) holes punched in it; it covers the hive entrance, allowing only bees through. A queen excluder at the top and bottom of stacks of stored equipment will also keep rodents out.

Wasps, Hornets and Robber Bees

These insect robbers can be a real menace, especially in late summer, but they are only looking for food. If your hive is bee-tight, it will also be proofed against other robbing insects. As well as making sure all boxes and roofs fit snugly, check that the ventilation holes in the roof are covered with fine gauze. Reduce the entrance size so the colony can defend itself easily. It is better for bees to queue up to go in and out than for wasps to sneak in round a corner.

Birds

Birds, especially swallows and tits, eat insects, including bees. Fortunately, numbers lost are generally very small and colonies can replace them. However, the green woodpecker can cause

A wasp is attacked by guard bees

considerable problems by pecking huge holes in the sides of boxes and eating bees, wax and honey. A beehive is like a hollow tree with a large number of insects inside, which is too great a temptation for the green woodpecker.

Did You Know?

'Jam traps' attract wasps into a bottle of water baited with jam. The narrow bottle entrance and smooth glass sides mean the wasps cannot get out and they drown.

Woodpecker Protection

The best protection is to wrap hives in small-mesh (25 mm/1 inch) chicken wire, enough to cover the whole depth of the box to the floor and to fold over the roof to keep it in place (*see* page 143). Treated carefully, it will last for years. A less expensive deterrent is to attach strips of heavy-duty plastic to the roof, making sure these hang down to the floor. They will flap in the wind to scare off the birds. Cut the strips slightly shorter over the entrance so the bees can still fly in and out.

217

Queen Problems

Under normal circumstances, the queen is the only female in the colony that can lay eggs, so its continued existence depends on her remaining vigorous and healthy. If the queen is lost or damaged, colonies can become weak or even die unless a new queen is produced or introduced quickly.

Failing and Drone-laying Queens

Occasionally, a queen runs out of sperm. She can then only lay unfertilized eggs that become drones. Similarly, if a virgin queen received inadequate sperm on her mating flight, her eggs will be unfertilized and she becomes a 'drone-laying queen'. Under these circumstances no new worker bees can be produced, making it impossible to raise a new queen, so the colony dwindles and eventually dies.

Queen Loss

The queen normally keeps herself safely protected inside the hive. However, if she dies or is lost, the colony is alerted rapidly by the change in pheromones. After 10–24 hours, the bees know they are queenless and start constructing 'emergency' queen cells. If worker eggs or very young larvae are present, the bees can raise a new queen. However, without this young brood, the colony will become hopelessly queenless and is doomed unless the beekeeper intervenes.

Laying Workers

A hopelessly queenless colony may develop laying workers. Female workers have rudimentary ovaries but are normally inhibited from laying eggs by pheromones from both the queen and the brood. Without these, that restraint is removed and a few workers will start laying eggs. However, they cannot mate so can only lay unfertilized eggs which become drones, again signalling the colony's ultimate demise.

Recognizing the Problem

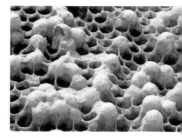

These conditions are recognized by carefully inspecting the brood combs. Drone-laying queens and laying workers both produce irregular patches of drone brood laid in worker cells. The combs become uneven, with scattered groups of drone cells and distorted cappings as the bees try to accommodate the larger drone larvae in the smaller worker cells. Laying workers do not lay consistently like the queen would, and often place eggs on the cell walls because their bodies cannot reach the bottom. They may also lay multiple eggs in a cell.

Drone brood laid
by workers

Rectifying the Problem

First carefully check whether or not the queen is still there. As long as they have sufficient adult bees, strong colonies with drone-laying queens can be saved by finding and killing the old queen and either introducing a new queen or uniting the colony with another. However, colonies containing laying workers are very difficult to retrieve, with the bees usually killing any new queen that is introduced. It is best to regard the colony as lost, but the bees do not have to be destroyed. Take each frame in turn and shake the bees onto the ground in front of another strong colony. Remove the hive of the laying worker colony from the apiary. The remaining workers will find their way into the other colony.

Disease Prevention Strategies

Vigilance and common sense will help avoid major disease problems. Brood diseases are spread between colonies by any means whereby infected bees, honey or comb cross from colony to colony. The beekeeper is the main culprit, closely followed by robbing. Careful apiary design, prevention of robbing and proper apiary hygiene will minimize disease risks. Always check colonies for brood disease before moving combs between them.

It is important to learn how to check for disease – advice from experienced beekeepers is best

Healthy sealed and unsealed brood

Minimizing Disease Risks

▶ Make regular health checks part of your routine inspections

▶ Learn to recognize what is healthy and ask for advice if worried

▶ Use your local bee inspector as a resource

▶ Only exchange frames between colonies if you are sure they are disease-free

▶ Never feed bees honey from bees other than your own, and ideally never feed honey

▶ Avoid anything that causes robbing

▶ Never leave supers or frames out for bees to rob

▶ Return wet combs to their own colony

▶ Practise good apiary hygiene

▶ Keep tools, suits, gloves and other equipment clean and sterilized

▶ Arrange the apiary to minimize drifting

▶ If a colony appears not to be thriving without reason, check for brood disease

▶ If uniting a weak colony with a strong one, first determine why it is weak – if it is diseased, you could spoil a good colony with a bad one

▶ Ideally, supers would always be used on the same hive but at least try to keep them in the same apiary

▶ Super combs used for brood rearing are more likely to be swapped between hives, together with any unseen infection they may carry

▶ If a colony dies for an unknown reason, close up the hive and call the bee inspector

▶ Inspect for EFB only when plenty of unsealed brood is present

▶ AFB spores can remain viable virtually indefinitely; dried scales are easily missed on old, blackened comb

▶ Sterilize used equipment and avoid second-hand comb or combs from unknown sources

▶ Migratory beekeeping and sales of bees can spread diseases widely

▶ Collect stray swarms from unknown sources so they cannot form pockets of infection elsewhere, and keep them isolated until you are sure they are pest and disease-free

▶ Avoid stress for bees, such as unnecessary handling and prolonged periods of movement or closure

▶ Minimize the risk of brood disease by hiving swarms onto foundation and not feeding immediately; the bees use any infected honey for wax production which locks spores away from harm

▶ Isolate (quarantine) brought-in bees for 6–8 weeks and inspect the brood thoroughly before moving them to the apiary. Of course, if you are a beginner buying in one colony (or collecting a single swarm), you will be effectively isolating them anyway.

▶ Use acetic acid to fumigate all stored combs

▶ Clean and sterilize hives and other equipment before storage

▶ Replace old brood combs with frames of foundation regularly and systematically

▶ Consider shook-swarm comb changes, especially in areas at high risk of EFB

▶ Keep your BDI insurance up-to-date and sufficient for your colony numbers

Key Points in Beekeeping Hygiene

▶ Do not leave bits of comb or honey lying around in the apiary

▶ Collect brace comb or scrapings from the top bars and excluder in a covered container

▶ Recycle the wax, most easily done in a solar wax extractor

▶ Keep your hive tool clean to reduce the risk of disease transmission

▶ Keep equipment clean with appropriate disinfectants

▶ Wear rubber gloves when using washing soda solution to clean and disinfect hive tools and smokers

▶ Scrub smokers clean with soapy water or washing soda solution

▶ Disinfect second-hand hives before use; scrape the boxes and burn the scrapings; scorch wooden parts with a blowlamp until the timber darkens but do not burn it

▶ Disinfect plastic components using a solution of a household bleach containing 3 per cent sodium hypochlorite

▶ Fumigate combs with acetic acid

▶ Wash overalls regularly, adding a small quantity of washing soda crystals with the detergent to help remove propolis

▶ Wash veils by hand to avoid damage

Fumigating Combs

Comb fumigation with acetic acid (80 per cent solution) is highly effective against EFB, chalk brood, Nosema and certain stages in the wax moth lifecycle. Combs to be sterilized must be clean, as acetic acid does not penetrate organic matter or honey residues. Place a good wad of cotton wool in a honey jar lid and add about 50 ml of acid for every two supers. Place a sheet of newspaper between every two supers to inhibit movement of adult moths and ensure fumes permeate the whole stack. Cover the stack with a roof or travelling screen to stop mice and other pests getting in.

Health and Safety

Acetic acid is corrosive, so do not breathe in any fumes. Clean up splashes or spills immediately using plenty of water. In the case of contact with eyes or skin, wash immediately with plenty of water and seek medical attention urgently. When diluting acids, always add the acid to the water. Store chemicals securely in clearly labelled bottles with the purchase or use-by date. Keep them well away from children. Dispose of any unwanted chemicals in an environmentally responsible way. Acetic acid will corrode metal hive parts.

Shook-swarm Comb Changes

For diseases such as EFB, where a significant residue of the pathogen remains in the combs, changing them at an appropriate time of year will give the colony a better chance of overcoming the problem. Shake bees from their combs into a new sterilized hive containing frames of foundation and then feed with sugar syrup. Comb changes should only be undertaken in late spring/early summer, depending on the season and locality, to give the colony sufficient time to draw out the comb and build up to full strength. This technique is most suitable for strong colonies. Colonies must be fed heavily so they can draw out new comb, especially if the weather is poor.

What You'll Need

▶ A clean brood box into which to transfer the colony

▶ A second box for temporary storage of the frames

▶ A full set of frames fitted with foundation

▶ A queen excluder

▶ A queen cage

▶ A clean floor

▶ A contact feeder

▶ 2.5 litres (4⅓ pints) of 2:1 sugar syrup

Top Tip

Colonies treated by shook swarming often become the strongest and most productive in an apiary. This is because Nosema, chalk brood and Varroa mites have also been cleared out of the colony.

The Shook Swarm Method

▶ Prepare a clean brood chamber, filled with frames of foundation, a floor, crownboard and queen excluder

▶ Move the colony to be transferred to one side and place the clean floor on the original site

▶ Place a queen excluder on top to prevent the queen leaving the hive

▶ Add the clean brood box with frames of foundation; the flying bees will return to this box

▶ Remove three or four frames from the centre, making a gap into which to shake the bees

▶ Find the queen and put her somewhere safe so she can be released into the new chamber after the bees have been shaken in

▶ Hold each frame in turn well down in the gap in the new box and shake it sharply to dislodge all the bees. Any clinging to the comb can be brushed off

▶ Put the cleared frames into the spare box, covering it to prevent bees gaining access

▶ Knock or brush any bees clinging to the original brood box or other hive parts into the new one

▶ Replace the frames of foundation removed; rest them on the bees and gently ease them into position as bees climb onto them

▶ Release the queen back into the colony

▶ Add a contact feeder with 2.5 litres (4⅓ pints) of strong sugar syrup to help the bees draw out the foundation. In the case of control for EFB or other disease, delay feeding for two days so that any contaminated nectar carried by the bees is used in comb-building

▶ Close up the hive

▶ After about a week, once brood is present, remove the queen excluder

▶ Maintain feeding until all combs are drawn out unless there is a continuing nectar flow

Prepared gap in a box of new foundation

Shaking in the last bees

Key Points about Robbing

▶ Robber bees can steal another colony's honey stores, leaving it to starve; weak hives are at particular risk

▶ Robbing spreads pests and diseases

▶ Robbing is most often associated with the honey harvest in autumn

▶ Avoid robbing by preventing it from starting

▶ Do not put out cappings, supers or buckets with spare old honey for the bees to rob

▶ Reduce the entrances at risky times (that is, when taking off honey), and for weak colonies

▶ Never allow bees access to supers during honey extraction

▶ Begin sugar feeding in the evening

▶ If robbing has started, reduce all the entrances, especially on weak colonies; consider moving any especially important or vulnerable stocks to another location temporarily

▶ Keep equipment in good order

Returning the queen

Reduce the entrance to minimize robbing

Key Points to Prevent Colony Losses

▶ Winter survival depends on good autumn preparation and winter vigilance

▶ The biggest cause of winter losses is lack of food; bees need 16–18 kg (35–40 lb) of winter stores; hefting the hive occasionally during the winter, and especially in early spring when stores are being used rapidly, indicates how much food remains. Ideally, bees should always have stores above the cluster

▶ Feed candy (available from most baker's shops) if you think a colony needs food

▶ Small colonies are less likely to survive the winter than large ones; to winter easily, a colony should cover at least five brood frames

▶ Consider uniting small colonies but check why they are small; uniting a diseased colony with a healthy one will not improve colony survival

▶ Use good husbandry to keep bees healthy; effective monitoring and control of Varroa in particular are essential

▶ Ensure colonies are safe from adverse weather conditions, predators, robbing insects and vandals

▶ Ideally, keep no more than 10 colonies per apiary

Help on Disease

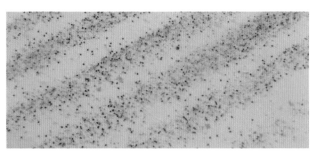

Beekeepers in England and Wales are very fortunate to have the services of the National Bee Unit (NBU), which is part of the Food and Environment Research Agency (Fera), based at Sand Hutton near York. The NBU operates a free disease-inspection service.

Monitor Varroa mite drop in the fight to prevent colony loss

Regions of England and Wales each have a full-time Regional Bee Inspector (RBI), supported by a team of Seasonal Bee Inspectors (SBIs) who will come and check your colonies for the notifiable bee diseases on request. All RBIs and SBIs are experienced beekeepers very willing to offer help and advice on all aspects of beekeeping, not just diseases.

Information available from the NBU

The NBU produces excellent descriptive leaflets on bee diseases which can be downloaded from its website, www.nationalbeeunit.com, regardless of where you live. The website also contains much useful information on diseases. If you live in England or Wales, you are very strongly advised to register on BeeBase (https://secure.fera.defra.gov.uk/beebase/public/Contacts/contacts.cfm). By knowing where beekeepers and their colonies are, the NBU is best placed to help all of us fight disease and keep our colonies healthy.

Diagnosing Diseases in the US

The Agricultural Research Service department of the United States Department of Agriculture (USDA) operates a Bee Disease Diagnosis Service for beekeepers across the US, for which there no charge.

Checklist

▶ **Know the signs:** It is important to be able to recognize what a healthy colony looks like before you can spot the signs of disease and problems.

▶ **The main threat:** *Varroa destructor* is the primary cause of colony losses.

▶ **Notifiable diseases:** Be aware of which diseases are notifiable by law.

▶ **Symptoms of AFB and EFB:** AFB causes dark, sunken cappings and the infected larva 'ropes'. EFB-infected larvae look 'melted' and die in awkward positions in the cells.

▶ **Symptoms of minor brood diseases:** Learn to tell the difference between relatively minor diseases, such as chalk brood (chalky white 'mummies') and sac brood ('Chinese slipper' sacs), and the more serious brood diseases.

▶ **Stay vigilant:** Colonies must be monitored for Varroa levels and treated accordingly. This will also help control potentially fatal viruses.

▶ **Don't delay treatment:** Treatment can be with varroacides or biotechnical methods.

▶ **Other pests:** Major pests include wax moths, rodents, wasps, hornets and green woodpeckers.

▶ **Adult diseases:** Varroa, Nosema, Amoeba and Acarine are the major adult bee diseases.

▶ **Terminal:** A drone-laying queen and laying workers signal the end of the colony.

▶ **Reduce your risk:** Good apiary hygiene minimizes disease risks. The shook swarm technique can be used to transfer a colony to clean comb.

▶ **Professional advice:** Take advantage of services such as free inspection.

229

Taking Things Further

Queen of the Castle

This final chapter will look at some of the finer skills needed to develop your beekeeping further. It will not cover the area of adding value to bee products (candle-making and the like) simply because of the limitations of space. What we will be concentrating on here is explaining more about the queen: how to find her, how to assess the characteristics of a colony (determined by the queen) and how to control swarming.

Book Learning

Of all the once-rural arts and crafts, beekeeping is one that is difficult to learn from a book. I can clearly remember my own confusion when faced with an exploded diagram of a beehive in my first beekeeping book. It made no sense at all until I saw it in real life.

Beekeeping Mentor

If possible, find a beekeeping mentor, preferably someone with both a reasonable length of practical experience and some beekeeping qualifications. And yes, it is possible to gain qualifications in beekeeping. Taking beekeeping assessments helps you measure your progress and understanding, as well as encouraging you to read widely and ask searching questions based on your own observations.

Finding the Queen

One of the most daunting tasks, especially for new beekeepers, can be to find the queen. It is a time-consuming activity and, under normal circumstances, not really necessary. As long as you can see eggs in worker cells, she is almost certainly there, but seeing her is always reassuring.

Concentrate

To find the queen you'll need to concentrate. You cannot combine this with any other tasks. It helps to think about the way the queen behaves and to have a picture of her in your mind. Although she is only slightly larger than a worker, the queen has a longer abdomen and legs. She will always avoid the light and run into the darkest part of the colony to keep safe. Before you disturbed her, she was busy laying eggs at the heart of the brood nest. Work quickly and smoothly with as little smoke as possible, arranging the colony so she cannot slip past you onto combs you have already checked.

First Find Your Queen

▶ Remove the first frame quickly but carefully; it is probably full of food so she is unlikely to be on it; scan it quickly and put it into a suitable box in the dark

▶ Look at the face of the comb that was in the dark first

▶ Do the same with the next comb, taking two frames from the brood box to leave a big gap; this lets in lots of light to discourage the queen from moving onto combs you have already checked

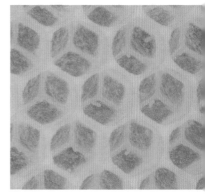

Eggs in worker cells indicate the presence of the queen

A queen surrounded by workers

233

▶ Check each comb systematically; scan the edges of the comb first and work toward the comb centre so she doesn't run onto the back of the comb while you are looking at the front

▶ Remember she might just be on the hive floor or body

▶ If you didn't find the queen, look again at the two combs in the box, before going through the colony a second time

▶ She is easiest to find in spring

Top Tip

Finding the queen can be difficult. Mentally picturing her and taking a positive attitude will help you find her.

If All Else Fails

▶ Arrange the combs in groups of two with gaps between; cover everything up, go and have a cup of tea then return and try again – she will be hiding between one of the pairs of combs

▶ Try again another day; two looks are enough

▶ Have someone looking over your shoulder; they seem to see her more quickly

Marking a queen

▶ Mark the queen when you find her (by picking her up and carefully sticking a tiny blob of coloured enamel paint on her thorax – or you can use a press-in queen cage, also known as a Baldock or 'crown of thorns', to restrain her and use a proper queen marking 'pen')

Testing for Queenlessness

To test if the queen is present, exchange a comb from the suspected queenless hive with one containing eggs and young larvae from another hive. If the colony is queenless, the workers will quickly start building queen cells to replace the lost queen. If the queen is still there, the brood will be allowed to develop normally. Of course, if a failing queen or laying workers are present in the colony, no queen cells will be built, but this is a rare occurrence.

Swarming, Prevention and Control

We have already considered collecting swarms as an economical method of gaining bees, a disease control mechanism and a public service. However, in your own bees, you want to prevent and control swarming. In general, this topic can confuse even quite experienced beekeepers. Many beekeepers worry that swarm control is complicated, but it is not difficult and it is entirely possible to stop colonies swarming.

Managing Nature

The biology of swarming has already been outlined, so here we will look at managing this natural reproductive process. The establishment of a new colony takes an enormous investment of the energy and resources of the parent colony, leading to a loss of the honey crop and the outside risk that either one or both parts of the divided colony dies out. On the plus side, a colony that has swarmed gains a vigorous new queen.

Swarm Prevention and Control
▶ Swarm prevention covers activities designed to delay or prevent the initiation of swarming
▶ Swarm control covers activities after a colony has started to make preparations for swarming

235

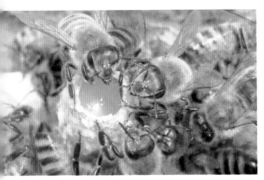
Workers building up a queen cell containing a larva bathed in royal jelly

Why Colonies Swarm

The queen substance pheromones are essential to keep colonies together as a unit and to suppress the urge of the workers to build queen cells. Colonies prepare to swarm in response to lack of queen pheromone. Overcrowding and poor ventilation will hinder the distribution of queen substance around the colony, while an ageing or poorly mated queen will produce insufficient queen substance. In either case, the consequence will be that the workers will start to build queen cells.

Swarm Prevention

Young, vigorous queens rarely swarm as the queen substance they generate is more than enough to keep the colony together. To prevent swarming, you need to control the factors that promote it.

Congestion

If the colony is very populous and the hive crowded, the airflow through it will be impeded, interfering with the distribution of the queen substance to the workers. Giving plenty of super space helps to reduce congestion in the hive and improve ventilation, allowing better distribution of queen substance. Plentiful foraging opportunities also reduce congestion because foraging bees are away from the hive. Clogged-up queen excluders can also encourage colonies to swarm.

Top Tip

Some races and strains of bees are more inclined to swarm than others. This characteristic is especially common in Carniolan bees.

Summary of Prevention Strategies

- Once your colony starts producing drones, inspect the brood nest regularly, ideally weekly
- You are looking for signs of swarming preparations, in the form of queen cells containing eggs or developing larvae
- Give the colony enough space by adding the next super when bees have reached the outside combs of the previous one
- Open entrances, clean wire excluders and open mesh floors all improve ventilation
- If you cannot decide whether or not another super is needed, give it to them anyway
- Giving space and wax-making/comb-building work helps to reduce the swarming impulse
- Keep stocks headed by one- or two-year-old queens
- Do not make new colonies from queens that swarm every year

Scraping the queen excluder – a clogged-up queen excluder can encourage swarming

Queen cells are often built at the bottom of a frame

Removing a queen cell

Queen Cells Indicate Swarming Preparations

Probably the easiest visible indication of swarming is the construction of queen cells. Little queen cups are common and do not in themselves indicate swarming. However, once the queen has laid an egg in a queen cup the workers will elongate the sides to accommodate the growing larva, and the swarming process has started.

Queen cells are unmistakable. When fully developed, they are about the size of an unshelled peanut with a similar surface texture, and hang down from the face of the comb. They can also be built at the sides and bottom of the comb. There may be 20 or more.

Once there are occupied queen cells and the swarming process has begun, the chances of a swarm emerging are quite high. However, the bees can change their minds and tear down the cells if conditions change.

Top Tip

Construction of queen cell cups does not indicate swarming preparations until they are occupied by an egg or larva. These 'play cups' are best pinched out.

238

Swarm Control

Once unsealed queen cells are found, prompt action is needed before the swarm leaves. The queen cell is sealed on the eighth day after the egg has been laid in a queen cell cup, and the virgin queen emerges on the sixteenth day. What usually surprises people is that the old queen leaves with the swarm as soon as the new queen cell(s) is sealed, so things can happen very quickly.

There are almost as many swarm control methods as there are beekeepers, some requiring elaborate pieces of equipment. However, it does not have to be complicated. Essentially, all swarm control methods act on the same basic principle, which is to make the bees 'think' they have swarmed already.

A queen emerging

Natural Swarms

A natural swarm is composed of the old queen with about half of the worker bees, usually those that are old enough to be foragers (and who know about the world outside the hive). The bees that remain are the younger bees looking after developing queen cells, one of which will produce their new queen.

Control Through Artificial Swarms

To control the swarm, the beekeeper needs to establish the natural situation before the swarm actually leaves. The flying bees (and the old queen) are separated from the brood (and the developing queen cells) and the colony is divided into two parts, which is what would happen naturally. If an extra colony is not needed, the two parts can be reunited later on in the year. We will look at one method where you 'first find the queen' – and one where you don't.

239

The Artificial Swarm Method

An artificial swarm is made if eggs or larvae are found in queen cells. An extra hive, complete with a full complement of frames, to house the new colony is needed.

1   Lift the hive containing the swarming colony to one side
2   Place a new floor, brood box and frames on the original site; the flying bees will come back here
3   Find the queen in the swarming colony
4   Place the queen, on her frame with the adhering bees, into the new brood box on the original site
5   Fill up the new brood box with frames, preferably of drawn comb
6   Add the queen excluder and any supers, then close this hive
7   Move the hive containing the brood and young bees at least six feet away to a permanent stand elsewhere in the apiary
8   Reduce the number of queen cells in this section to just a couple; select the biggest and best, removing any that are misshapen, small or already sealed
9   Replace the frame removed with the queen to fill up the box
10 Add the crownboard and roof to close the second hive
11 Feed both parts as needed

As a slightly more sophisticated option that confers some extra benefits, you can modify the procedure as follows.

▶   **At step 7:** Rather than move the swarming hive (containing the brood) to its new position immediately, position it to one side of the original site with its entrance at 90° to its original orientation.

> ## Top Tip
> Most methods of swarm control involve using extra equipment, including frames with foundation or drawn comb, which need to be ready in advance.

> ## Top Tip
> A colony can be regarded as having three parts: the queen, the brood and the flying bees. All swarm-control methods separate one of these from the other two.

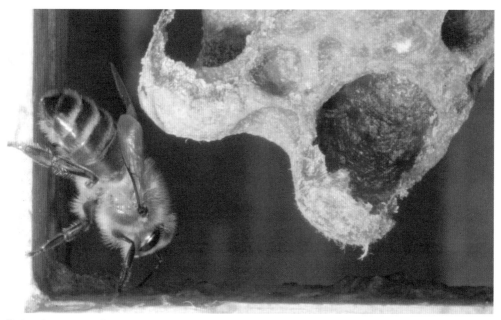

Broken-down queen cells

▶ **Seven days later:** Move the hive at least six feet away to its permanent position elsewhere in the apiary; the bees that started foraging during the week will have learnt to return to the hive in its first position; because their hive has gone they will join the hive on the original site.

The colony with the brood, now in its second position, will become even more depleted of bees; depleting it of bees discourages any possible after-swarms; instead the first virgin will kill any others and the house bees will remove any remaining queen cells

Swarm Control without Finding the Queen

There will be times when you simply cannot find the queen. All is not lost. It is still possible to control swarming. This method depends on knowing that the nurse bees will never leave brood unattended. Every single bee is systematically brushed into a new brood box containing a full set of combs. By default they must include the queen. The queen excluder is replaced and the now beeless brood is put back above the queen excluder. The nurse bees scramble though the excluder to continue their job of keeping the brood warm, thereby separating the flying bees and queen from the brood and developing queen cell(s).

The Method – Day One

1   Lift the brood box containing the elusive queen to one side (leaving the floor in place)
2   Put a new brood box, with a full complement of frames, onto the original floor on the original site
3   Remove six or seven frames from the centre leaving a large gap

4   Take each comb from the original brood box in turn; hold it well down inside the new brood box and gently shake or brush all the bees from the comb into the new box; the queen will be in one of three places – among the bees in the new box, lurking on the original floor or on the side of the old brood box

5   You do not want bees repopulating the combs you have cleared, so place them in a spare brood box and cover the top; keep the frames in the same order as they came out of the original brood box

6   Brush any bees remaining on the sides of original brood box into the new one; every single bee including the queen will now be in the new brood box

7   Replace the queen excluder

8   Place the brood box containing all the brood on top

9   Add the crownboard, then any super(s), and close up the hive

Two Broods are Born

Over the next 24 hours, the nurse bees will be attracted back to the brood in the upper brood box, leaving the queen behind because she cannot get through the excluder. The flying bees will have already gone back to work and will remain in the bottom brood box filling up the empty combs. All that remains is to divide the hive into two parts. This must be done the next day.

The Method – Day Two

10   Set up a new floor in the place you want to site the new colony

11   Move the top brood box containing all the brood and the nurse bees onto the new floor; the young bees will stay put because they have never been out of the hive

12   Reduce the number of queen cells to just a couple, selecting the biggest and best and removing any that are misshapen, small or already sealed

13   Share the supers between the two colonies

14   Reduce the entrance to a size that can be defended easily

15   Close up both hives as usual

16   Feed both parts as needed

243

Frustration Methods of Swarm Control

Some people may advocate using 'frustration' methods of swarm control. The main one among these is the simple removal of queen cells. This can sometimes work. However, I consider that if the bees are making swarm preparation they are sending the message that the queen is not up to scratch. It is very easy to make a colony queenless by removing queen cells. If it has already swarmed and you have not recognized this and removed all the queen cells, there is no way the bees can raise another queen. It is not to be recommended.

Problems Arising

Winnie the Pooh said the most sensible thing about beekeeping when he said,'You never can tell with bees'. So I need to mention a couple of things, because things may not always go entirely to plan.

Sealed Queen Cells

If some of the queen cells are sealed, it is likely that the colony has already swarmed. It is quite easy for this to happen unnoticed as the remaining colony settles down very quickly after swarming. The main clue that the swarm has left is that the number of bees in the colony has significantly reduced. It is still possible to divide the colony to make an increase or to prevent after-swarms, in the ways previously described, but in this case both parts need to have a queen cell. It takes a month or more for a new queen to hatch, mature, mate and get the colony growing again, so there will be a long period where divided or swarmed colonies will need your careful attention.

Top Tip
Once the possibility of swarming is over, weekly examinations are no longer necessary.

Scraping clean the top bars to ensure a snug fit, prior to uniting colonies

Uniting Colonies

If the queen does not mate properly, gets lost on her mating flight or you simply do not want any more colonies, uniting colonies over newspaper (*see* page 157) is a straightforward and effective procedure. Colonies will not normally join together without fighting, killing many worker bees. Each colony has a distinctive odour that ensures the worker bees know which is their home and when intruders are trying to invade it. To unite two colonies successfully, this odour has to be mixed in a way that prevents the bees from fighting.

When to Unite

A good time to do this is after the summer honey season, once the honey is harvested but before feeding and Varroa treatments start. If you wish to keep a particular queen, kill the queen you do not want, rather than leaving the two queens to fight it out. Your records, or maybe a gut feeling about which colony is better, will tell you which queen you want to head a colony in your apiary next year.

245

Uniting over Newspaper

▶ Kill the queen you do not want, or allow the bees to choose if you do not have a preference

▶ Clean the tops and bottoms of the frames, if necessary, so the two brood boxes fit neatly together

▶ Cover the whole of the bottom brood box area with newspaper

▶ Place the queenless, smaller or moved colony on top

▶ Close up the colony

▶ Reorganize the frames the following week: put all the combs containing brood in the bottom brood box and any others in the top one; position food combs over the brood nest

▶ The bees can overwinter like this and the colony can be reduced to a single brood box in the spring

General Rules for Uniting

▶ When uniting a 'queenright' colony with a queenless one, the queenless colony goes on top

▶ When uniting colonies of different sizes, such as uniting a nucleus to a full colony to requeen it, the smaller colony (nucleus) goes on top

▶ If you are uniting colonies within the apiary, the colony you have moved goes on top

Preventing Swarming – Summary

▶ Give the colony plenty of space to help prevent swarms

▶ Inspect the brood nest regularly

▶ Decide your swarm control method in advance

▶ Have enough equipment ready

▶ If you do not wish to increase colony numbers, or the colonies are small, unite them before winter

What Makes a Good Queen

A young, vigorous queen is central to good beekeeping. However, like people, not all queens have the same nature. The queen gives the colony its characteristics, be it docility, a strong instinct to protect their nest, a propensity to swarm, or the ability to overwinter with thrifty use of stores. In practice, not all the desired characteristics are likely to be available in a single queen, but it is a good thing for beekeepers to be discriminating when assessing their colonies.

What Characters Do You Want?

This is, of course, a personal choice and beekeepers may add details such as white cappings for prize-winning comb honey. However, here are some of the accepted characteristics that beekeepers look for in a queen (and thus in the colony).

Desirable Characteristics

▶ Good temper or docility – the bees are gentle to handle

▶ Non-following – the bees stay in the apiary when you leave

▶ Good honey-getting

▶ Thrifty food consumption and the capacity to adapt consumption to available forage

▶ Good overwintering ability

▶ A low propensity to swarm or a tendency to supersede

▶ Stillness on the comb, making queen-finding easier

▶ Disease resistance

How Do You Get Better Queens?

The two characteristics that make good queens are their genetic inheritance and how well they are nourished during their development in the queen cell. The latter affects the numbers of eggs queens are able to produce in their lifetime.

Top Tip

Keep records of all your colony inspections so that you can make an informed assessment at the end of the season and plan for the next.

Did You Know?

The best bees will undoubtedly be found in your own neighbourhood and you can improve these by being selective about the colonies you keep.

Buying Bees

As a novice beekeeper, you can only really trust to luck when buying colonies. Buying them from reputable source, especially a local one, will help. Starting with a swarm may mean you get lumbered with a very swarmy strain of bee, which will take a lot of time and attention. However, changing the queen completely changes the nature of the colony, often within weeks. Queens can be purchased and introduced into a new nucleus hive or, as experience grows, they can be reared and even bred for desired characteristics.

How Do I Breed Better Queens?

Breeding bees, as with any animals, requires getting the best queens to mate with the best drones. The best queens and drones are selected by keeping relevant records of their performance. Breeding is not straightforward because it is difficult to control the drones with which the queen mates. I do not propose to go into it further here as there are many specialist books on the topic.

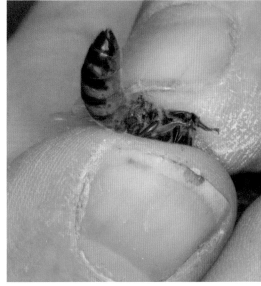

Killing queens is sometimes necessary

Summary
- Look for desired characteristics
- Keep careful records
- Learn how to rear queens
- Find a way to control mating if you can
- Cull queens that are not up to scratch

249

Queen Rearing

Queen rearing can be as simple as utilizing the queen cells developed under the swarming impulse to create new colonies, or as complicated as using specially designated colonies as breeding colonies, cell-raising colonies and cell-finishing colonies. In general, people have two objectives when rearing queens. The first is to increase the number of colonies they own. The second is as part of a bee-breeding programme designed to change or protect the genetic characteristics of the colony.

Common Principles

There are many different methods of rearing new queens, but the underlying principles are very similar to those we have outlined for swarm control. For the bees to rear new queens they must be queenless, or they must think they are. Consequently, if the queen is removed from the colony, the bees will start to make new queens. For a colony to rear satisfactory queens, whether naturally or as part of a queen-rearing programme, there must be fertile eggs and plenty of nurse bees present in a very strong and well-fed colony. The best queen cells are produced when the conditions are as near as possible to the natural conditions found when a strong colony of bees swarms.

Carry on Learning

Once you are comfortable with the basic ideas of beekeeping described here and understand the biology of the bees, then queen rearing is a useful skill to learn. There is just so much to know about bees, it can keep you finding out something new for a lifetime. Whatever your interest, there will be something for you. I do hope you will enjoy your beekeeping career.

HAPPY BEEKEEPING!

Checklist

▶ **Learn from experience:** A beekeeping mentor will help you learn more quickly.

▶ **Think positive:** Concentrating and taking a positive attitude helps when trying to find the queen.

▶ **Test for queenlessness:** This is done by introducing a comb of eggs and very young larvae.

▶ **Give them space:** Swarming can be prevented by giving a colony plenty of room.

▶ **Lack of queen substance:** Colonies prepare to swarm when the distribution of queen pheromone through the colony is insufficient.

▶ **Know the signs:** Swarming preparations can be recognized by the presence of eggs and larvae in queen cells.

▶ **Basic principle:** Swarm control hinges on separating one part from the other two.

▶ **Bees don't wait:** If you find sealed queen cells, your colony has almost certainly already swarmed.

▶ **Many methods:** Swarm control does not necessarily involve finding the queen.

▶ **Avoid fights between colonies:** Uniting colonies over newspaper is one of the easiest and most effective methods.

Websites and Further Reading

There are many useful websites and books on bees, beekeeping and associated subjects. I have tried to list a few in each area, but for a wider selection, contact specialist bee booksellers. Some of the books listed are out of print but it is worth trying to obtain them. Your local beekeeping association may well have copies in its library or you may be able to borrow them from your public library.

Websites

www.bbka.org.uk
Website of the British Beekeepers' Association with useful information about aspects of beekeeping. Includes contact details for local associations.

www.beedata.com
Northern Bee Books is a UK bookseller specializing in books about bees and beekeeping. It carries a large range of English-language bee books.

www.bee-craft.com
Website of the publisher of *Bee Craft*, with beekeeping information and a link for purchasing the journal, other publications and associated beekeeping items.

www.beeginners.info
Lots of useful and interesting information about bees and beekeeping.

www.bees-online.com
A comprehensive site giving further details on many aspects of beekeeping.

www.food.gov.uk
The website of the UK government's Food Standards Agency with information about legislation covering honey.

www.honey.com
Website of the US National Honey Board with comprehensive information about honey and also recipes.

www.honeyshop.co.uk
Bee Books New and Old stock books on beekeeping and associated subjects such as propolis, mead and other crafts.

www.ibra.org.uk
The International Bee Research Association is the world's longest-established publisher of apicultural research topics. It also has a comprehensive bookshop.

www.irishbeekeeping.ie
Website of the Federation of Irish Beekeepers including a useful list of bee forage.

www.nba.org.nz
Website of the New Zealand Beekeepers' Association, from which can be ordered *Elimination of American Foulbrood Disease without the Use of Drugs* and *Control of Varroa: A Guide for New Zealand Beekeepers*.

www.scottishbeekeepers.org.uk
Website of the Scottish Beekeepers' Association including a range of technical data sheets.

http://outdoorplace.org/beekeeping/citybees.htm
Gives notes on keeping bees in urban or suburban areas.

https://secure.fera.defra.gov.uk/beebase
BeeBase is the website of the National Bee Unit for England and Wales. It includes comprehensive information about bee pests and diseases. Beekeepers in England and Wales are strongly advised to register on BeeBase.

www.wbka.com
Website of the Welsh Beekeepers' Association including a growing list of useful articles.

Further Reading

Aston, David and Bucknall, Sally, *Plants and Honey Bees: Their Relationships*, Mytholmroyd, Northern Bee Books (2004)

Bee Craft Ltd, The Bee Craft Apiary Guides to Bee Diseases, Stoneleigh, Bee Craft Ltd (2005)

Bee Craft Ltd, The Bee Craft Apiary Guides to Integrated Pest Management, Stoneleigh, Bee Craft Ltd (2005)

Bee Craft Ltd, The Bee Craft Apiary Guides to Record Keeping, Stoneleigh, Bee Craft Ltd (2007)

Bee Craft Ltd, The Bee Craft Apiary Guides to Swarming and Swarm Control, Stoneleigh, Bee Craft Ltd (2007)

Bee Craft Ltd, The Bee Craft Apiary Guides to Colony Make-up, Stoneleigh, Bee Craft Ltd (2010)

Brown, R. H., Beeswax, 2nd ed., Burrowbridge, Bee Books New and Old (1989)

Collins, *Collins Beekeeper's Bible,* London, HarperCollins (2010)

Crane, Eva, *A Book of Honey,* Oxford, Oxford Paperbacks (1980)

Davis, Celia F., *The Honey Bee Inside Out,* Revised ed., Stoneleigh, Bee Craft Ltd (2006)

Davis, Celia F., *The Honey Bee Around and About,* Stoneleigh, Bee Craft Ltd (2007)

Flottum, Kim, *The Backyard Beekeeper - Revised and Updated, Dallas, TX, Quarry Press (2009)*

Flottum, Kim, *The Honey Handbook,* London, Apple Press (2009)

Furness, Clara, *How to Make Beeswax Candles,* reprint, Cardiff, International Bee Research Association (2010)

Gibb, *Andrew, Setting up and Managing an Apiary Site,* Stoneleigh, Bee Craft Ltd (2010)

Goodwin, Mark, *Eliminating American Foulbrood Disease without the use of Drugs,* Wellington, New Zealand Ministry of Agriculture and Forestry (2006)

Goodwin, Mark and Taylor, Michelle, *Control of Varroa: A Guide for New Zealand Beekeepers,* Revised ed., Wellington, New Zealand Ministry of Agriculture and Forestry (2007)

Gould, James L. and Gould, Carol Grant, *The Honey Bee,* New York, Scientific American Library (1988)

Hooper, Ted and Taylor, Mike, *The Bee Friendly Garden,* Yeovil, Somerset, Alphabet and Image Ltd (2006)

Morse, Roger A. and Flottum, Kim (eds), *Honey Bee Pests,* Predators and Diseases, 3rd ed., Medina, OH, The AI Root Co Ltd (1998)

Popescu, Charlotte, *The Honey Cookbook,* Pewsey, Wiltshire, Cavalier Paperbacks (1997)

Riches, Harry, *Insect Bites and Stings: a Guide to Prevention and Treatment,* Cardiff, International Bee Research Association (2003)

Riches, Harry, *Mead: Making, Exhibiting and Judging,* Burrowbridge, Bee Books New and Old (2009)

Schramm, Ken, *The Compleat Meadmaker,* Boulder, CO, Brewers Publications (2003)

Seeley, Thomas D., *The Wisdom of the Hive: The Social Physiology of Honey Bee Colonies,* Cambridge, MA, Harvard University Press (1995)

Seeley, Thomas D., *Honeybee Democracy,* Princeton, NJ, and Oxford, Princeton University Press (2010)

Tautz, Jürgen, *The Buzz about Bees,* Berlin, Springer-Verlag (2008)
Von Frisch, Karl, *The Dance Language*

and Orientation of Bees, Cambridge, MA, Harvard University Press (1967)

Waring, Adrian, Better *Beginnings for Beekeepers,* 2nd ed., Doncaster, BIBBA (2004)

Waring, Adrian and Waring, Claire, *Get Started in Beekeeping* (Teach Yourself), London, Hodder Education (2010)

White, Joyce and Rogers, *Valerie, Honey in the Kitchen,* rev. ed., Charlestown, Bee Books New and Old (2000)

White, Joyce and Rogers, Valerie, *More Honey in the Kitchen,* rev. ed., Charlestown, Bee Books New and Old (2001)

Williams, John, *Starting Out with Bees,* Stoneleigh, Bee Craft Ltd (2010)

Winston, Mark L., *The Biology of the Honey Bee,* Cambridge, MA, Harvard University Press (1987)

Beekeeping Journals

American Bee Journal, published monthly by Dadant and Sons Inc., 51 S. 2nd Street, Hamilton, IL 62341, USA

Bee Craft, published monthly by Bee Craft Ltd, 107 Church Street, Werrington, Peterborough PE4 6QF (www.bee-craft.com)

Bee Culture, published monthly by The A.I. Root Co, 623 W. Liberty St., Medina, OH 44256, USA

Beekeepers Quarterly, published quarterly by Northern Bee Books, Scout Bottom Farm, Mytholmroyd, Hebden Bridge, West Yorkshire HX7 5JS (www.beedata.com)

Bee Disease Insurance

Colonies can be insured against losses resulting from European Foul Brood and American Foul Brood with Bee Diseases Insurance Ltd, 57 Marfield Close, Walmley, Sutton Coldfield, West Midlands B76 1YD

253

Index